INCREASING CAPACITY FOR
STEWARDSHIP
OF OCEANS AND COASTS

A Priority for the 21st Century

Committee on International Capacity-Building for the
Protection and Sustainable Use of Oceans and Coasts

Ocean Studies Board

Division on Earth and Life Studies

NATIONAL RESEARCH COUNCIL
OF THE NATIONAL ACADEMIES

THE NATIONAL ACADEMIES PRESS
Washington, D.C.
www.nap.edu

THE NATIONAL ACADEMIES PRESS 500 Fifth Street, NW Washington, DC 20001
NOTICE: The project that is the subject of this report was approved by the Governing Board of the National Research Council, whose members are drawn from the councils of the National Academy of Sciences, the National Academy of Engineering, and the Institute of Medicine. The members of the committee responsible for the report were chosen for their special competences and with regard for appropriate balance.

This material is based on work supported by the Gordon and Betty Moore Foundation (Contract/Grant 611), the David and Lucile Packard Foundation (Contract/Grant 2005-28827), the National Oceanic and Atmospheric Administration (Contract/Grant DG133R04Q0009), the National Science Foundation (Contract/Grant OCE-0541186), the Marisla Foundation (Contract/Grant 3-07-142), and the Curtis and Edith Munson Foundation. Any opinions, findings, conclusions, or recommendations expressed in this material are those of the authors and do not necessarily reflect the views of the National Science Foundation, of the National Oceanic and Atmospheric Administration or any of its subagencies, or of the other organizations or agencies that provided support for the project.

International Standard Book Number-13: 978-0-309-11376-2 (Book)
International Standard Book Number-10: 0-309-11376-8 (Book)
International Standard Book Number-13: 978-0-309-11375-5 (PDF)
International Standard Book Number-10: 0-309-11375-X (PDF)
Library of Congress Control Number: 2008921427

Cover art created by Ernesto Reyes and Benjamín Flores, members of the art group of Amigos de San Lorenzo (Friends of San Lorenzo) in Achiote, Colón, Panamá. The art group is supported by the Centro de Estudios y Acción Social Panameño (Center of Panamanian Research and Social Action).

Additional copies of this report are available from the National Academies Press, 500 Fifth Street, NW, Lockbox 285, Washington, DC 20055; (800) 624-6242 or (202) 334-3313 (in the Washington metropolitan area); Internet, http://www.nap.edu.

THE NATIONAL ACADEMIES
Advisers to the Nation on Science, Engineering, and Medicine

The **National Academy of Sciences** is a private, nonprofit, self-perpetuating society of distinguished scholars engaged in scientific and engineering research, dedicated to the furtherance of science and technology and to their use for the general welfare. Upon the authority of the charter granted to it by the Congress in 1863, the Academy has a mandate that requires it to advise the federal government on scientific and technical matters. Dr. Ralph J. Cicerone is president of the National Academy of Sciences.

The **National Academy of Engineering** was established in 1964, under the charter of the National Academy of Sciences, as a parallel organization of outstanding engineers. It is autonomous in its administration and in the selection of its members, sharing with the National Academy of Sciences the responsibility for advising the federal government. The National Academy of Engineering also sponsors engineering programs aimed at meeting national needs, encourages education and research, and recognizes the superior achievements of engineers. Dr. Charles M. Vest is president of the National Academy of Engineering.

The **Institute of Medicine** was established in 1970 by the National Academy of Sciences to secure the services of eminent members of appropriate professions in the examination of policy matters pertaining to the health of the public. The Institute acts under the responsibility given to the National Academy of Sciences by its congressional charter to be an adviser to the federal government and, upon its own initiative, to identify issues of medical care, research, and education. Dr. Harvey V. Fineberg is president of the Institute of Medicine.

The **National Research Council** was organized by the National Academy of Sciences in 1916 to associate the broad community of science and technology with the Academy's purposes of furthering knowledge and advising the federal government. Functioning in accordance with general policies determined by the Academy, the Council has become the principal operating agency of both the National Academy of Sciences and the National Academy of Engineering in providing services to the government, the public, and the scientific and engineering communities. The Council is administered jointly by both Academies and the Institute of Medicine. Dr. Ralph J. Cicerone and Dr. Charles M. Vest are chair and vice chair, respectively, of the National Research Council.

www.national-academies.org

COMMITTEE ON INTERNATIONAL CAPACITY-BUILDING FOR THE PROTECTION AND SUSTAINABLE USE OF OCEANS AND COASTS

MARY (MISSY) H. FEELEY (*Cochair*), ExxonMobil Exploration Company, Houston, Texas
SILVIO C. PANTOJA (*Cochair*), University of Concepción, Chile
TUNDI AGARDY, Sound Seas, Bethesda, Maryland
JUAN CARLOS CASTILLA, Pontificia Universidad Católica de Chile, Santiago
STEPHEN C. FARBER, University of Pittsburgh (ret.), Santa Fe, New Mexico
INDUMATHIE V. HEWAWASAM, The World Bank, Washington, DC
JOANNA IBRAHIM, University of the West Indies, St. Augustine, Trinidad
JANE LUBCHENCO, Oregon State University, Corvallis
BONNIE J. MCCAY, Rutgers University, New Brunswick, New Jersey
NYAWIRA MUTHIGA, Wildlife Conservation Society, Mombasa, Kenya
STEPHEN B. OLSEN, University of Rhode Island, Narragansett
SHUBHA SATHYENDRANATH, Partnership for Observation of the Global Oceans, Dartmouth, Nova Scotia, Canada
MICHAEL P. SISSENWINE, National Oceanic and Atmospheric Administration (ret.), Woods Hole, Massachusetts
DANIEL O. SUMAN, University of Miami, Florida
GISELLE TAMAYO, University of Costa Rica, Heredia (*resigned May 2007*)

Staff

SUSAN ROBERTS, Director
FRANK R. HALL, Program Officer
JODI BOSTROM, Research Associate
NORMAN GROSSBLATT, Senior Editor

OCEAN STUDIES BOARD

SHIRLEY A. POMPONI (*Chair*), Harbor Branch Oceanographic Institution, Ft. Pierce, Florida

ROBERT G. BEA, University of California, Berkeley

DONALD F. BOESCH, University of Maryland Center for Environmental Science, Cambridge

JORGE E. CORREDOR, University of Puerto Rico, Lajas

KEITH R. CRIDDLE, University of Alaska Fairbanks, Juneau

MARY (MISSY) H. FEELEY, ExxonMobil Exploration Company, Houston, Texas

HOLLY GREENING, Tampa Bay National Estuary Program, St. Petersburg, Florida

DEBRA HERNANDEZ, Hernandez and Company, Isle of Palms, South Carolina

ROBERT A. HOLMAN, Oregon State University, Corvallis

CYNTHIA M. JONES, Old Dominion University, Norfolk, Virginia

KIHO KIM, American University, Washington, DC

WILLIAM A. KUPERMAN, Scripps Institution of Oceanography, La Jolla, California

ROBERT A. LAWSON, Science Applications International Corporation, San Diego, California

FRANK E. MULLER-KARGER, University of South Florida, St. Petersburg

JAY S. PEARLMAN, The Boeing Company, Kent, Washington

S. GEORGE H. PHILANDER, Princeton University, New Jersey

RAYMOND W. SCHMITT, Woods Hole Oceanographic Institution, Massachusetts

ANNE M. TREHU, Oregon State University, Corvallis

Staff

SUSAN ROBERTS, Director

SUSAN PARK, Program Officer

SHUBHA BANSKOTA, Financial Associate

PAMELA LEWIS, Administrative Coordinator

JODI BOSTROM, Research Associate

PREFACE

Ocean and coastal ecosystems are inextricably linked with humans. Nearly 40% of the world's population is concentrated in the 100-km-wide coastal zone of the continents, and many coastal residents in developing and developed countries depend directly on ocean and coastal ecosystems for their livelihood. Seafood is the primary source of protein for over a billion people, mostly in developing countries. The extraordinary natural productivity of ocean and coastal waters and the strategic benefits of a coastal location for trade, defense, industry, and food production have made oceans and coasts uniquely important.

The needs for the ocean and coastal ecosystems' goods and services are likely to increase substantially as the human population continues to grow, as more people move to coastal areas, and as people strive to improve their standard of living. As a consequence, the degradation of coastal and marine ecosystems is expected to worsen. That degradation necessitates the building of capacity, especially in developing countries, to ensure the future of ocean and coastal communities, the cultural heritage of indigenous peoples, and ecosystem-based services. Capacity-building for stewardship of the oceans and coasts is a complex multidimensional challenge and needs to be addressed as such. It requires interdisciplinary and multidisciplinary approaches to ensure that stakeholders develop the proper knowledge, skills, and attitudes to be effective stewards of the environment.

Capacity for ocean and coastal stewardship has been growing around the world as governments, development banks, donors, and the private sector have funded projects on various scales to address many issues. Those efforts have infused knowledge in people, transferred technology, and strengthened institutions. However, it is clear that the efforts need to be increased and be more effective in the future. Stewardship of ocean and coastal ecosystems that include people require highly interdisciplinary, flexible, and adaptive approaches to deal with transboundary issues in situations fraught with logistical, political, and conflict-resolution issues.

The study reported here was funded by the Gordon and Betty Moore Foundation, the President's Circle of the National Academies, the David and Lucile Packard Foundation, the National Oceanic and Atmospheric Administration, the National Science Foundation, the Marisla Foundation, and the Curtis and Edith Munson Foundation. To conduct this study, the National Research Council assembled a committee of international experts to examine current and past efforts in building the scientific, technological, and institutional capacities that countries need for developing and implementing effective ocean and coastal resource policies and to identify barriers to effective management that coastal nations encounter. This report presents the committee's deliberations and findings with respect to special challenges in achieving sustainable use of oceans and coasts, the evolution and limitations of past and current capacity-building, barriers to and constraints on effective capacity-building, and the way forward to increase capacity for effective governance and stewardship.

The committee is indebted to the staff of the Ocean Studies Board for their valuable services and willingness to work out complex meeting and workshop arrangements, obtain additional background material, and provide report editorial services. Frank Hall served as study director. His valuable insights, perspective, and lively sense of humor were much appreciated. We are especially indebted to Jodi Bostrom, who provided the day-to-day support of the committee and through her can-do attitude ensured that the deliberations of this international committee, spread across many time zones, were fruitful and constructive.

Mary (Missy) H. Feeley and Silvio C. Pantoja, *Committee Co-Chairs*

ACKNOWLEDGMENTS

This report represents the efforts of many individuals and organizations. The committee thanks Alan Sielen and the U.S. Environmental Protection Agency; they were instrumental in developing the conceptual basis for this study. While Mr. Sielen was a visiting scholar at the Ocean Studies Board, he worked with the U.S. Department of State to convene a symposium[1] in Washington, DC, on capacity-building for oceans and coasts that contributed greatly to the design and initiation of this study.

The report was greatly enhanced by the participants in the committee meetings and the workshop held as part of the study. The committee acknowledges the efforts of those who gave presentations at meetings and who submitted written statements: Glenis Binns, Genevieve Brighouse, Peter Burbridge, Loke-Ming Chou, Patrick Christie, Biliana Cicin-Sain, Clara Cohen, Harlan Cohen, Barry Costa-Pierce, Charlotte Elton, Mirei Endara de Heras, Julio Escobar, Henrik Franklin, Jeremy Harris, Marea Hatziolos, Lorna Inniss, Takashi Ito, David Kingsbury, Rosa Montañez, Orlando Osorio, Zuleika Pinzón, Ira Rubinoff, Kristin Sherwood, Richard Spinrad, Líder Sucre, Elizabeth Tirpak, Stella Maris Vallejo, Richard Volk, and Edward Urban, Jr. Their input helped to set the stage for fruitful discussions in the closed sessions that followed.

The committee also thanks the following for providing an enriching experience during the field trip to Achiote, Panamá: Orlando Acosta, Almyr Alba, José Angulo, Ana Isabel Araúz, Gilberto Barrio, Jeila de la Cruz, Julián de la Cruz, María Inés Díaz, Carlos Darinel Domínguez, Carlos Fitzgerald, Benjamín Flores, Carlos Gómez, Daniel Holness, Manuel Jaén, Felipe Martínez, Daniel Moreno, and Michelle Pobjoy.

This report has been reviewed in draft form by persons chosen for their diverse perspectives and technical expertise in accordance with procedures approved by the

---

[1]U.S. Department of State. 2004. Capacity Building for the Protection and Sustainable Use of Oceans and Coasts. Proceedings of a Symposium held November 8–9, 2004, in cooperation with the Ocean Studies Board of the National Academies, Washington, DC.

National Research Council's Report Review Committee. The purpose of this independent review is to provide candid and critical comments that will assist the institution in making its published report as sound as possible and to ensure that the report meets institutional standards of objectivity, evidence, and responsiveness to the study charge. The review comments and draft manuscript remain confidential to protect the integrity of the deliberative process. We thank the following for their participation in their review of this report:

MARGARET BOWMAN, Pew Charitable Trusts, Washington, DC
MARIA DE LOS ANGELES CARVAJAL, Conservation International (ret.), Sonora, México
EHRLICH DESA, Intergovernmental Oceanographic Commission, Paris, France
DAVID FLUHARTY, University of Washington, Seattle
HENRIK FRANKLIN, Inter-American Development Bank, Washington, DC
WILLIAM FREUDENBURG, University of California, Santa Barbara
LORNA INNISS, Government of Barbados, St. Michael
STEPHEN G. MONISMITH, Stanford University, California
SEBASTIAN TROENG, Conservation International, Washington, DC
KARL K. TUREKIAN, Yale University, New Haven, Connecticut
STELLA MARIS VALLEJO, United Nations Train-Sea-Coast Program (ret.), Cascais, Portugal
RICHARD VOLK, U.S. Agency for International Development, Washington, DC

Although the reviewers listed above have provided many constructive comments and suggestions, they were not asked to endorse the conclusions or recommendations, nor did they see the final draft of the report before its release. The review of this report was overseen by **Michael K. Orbach**, Duke University, Beaufort, North Carolina, appointed by the Divison on Earth and Life Studies, and **John E. Dowling**, Harvard University, Cambridge, Massachusetts, appointed by the Report Review Committee, who were responsible for making certain that an independent examination of this report was carried out in accordance with institutional procedures and that all review comments were carefully considered. Responsibility for the final content of this report rests entirely with the authoring committee and the National Research Council.

CONTENTS

Summary   1

1   Introduction   11
    Origin of the Study, 12
    Key Concepts, 13
    Report Organization, 15

2   The Challenges of Achieving Stewardship of Oceans and Coasts   16
    Ocean and Coastal Ecosystems and Services, 16
    Challenges to Our Ocean and Coastal Ecosystems, 19
    Moving Toward Ocean and Coastal Stewardship and
        Ecosystem-Based Management, 22
    Findings and Recommendations, 27

3   Growing Capacity for Stewardship of Oceans and Coasts: A Work in Progress   29
    Investors and Investments in Capacity-Building, 30
    How to Grow Capacity, 33
    Findings and Recommendations, 48

4   Moving Toward Effectiveness: Identifying Barriers to and Constraints
    on Effective Capacity-Building   50
    Barriers to and Constraints on Capacity-Building, 50
    Principles for Effective Capacity-Building, 60
    Findings and Recommendations, 61

5 What Aspects of Capacity-Building Need More Emphasis?   63
   Needs Assessments for Capacity-Building, 64
   Sustaining Capacity and Capacity-Building Efforts, 64
   The Need for Effective Program Assessment, 68
   Professional Standards, 70
   Information for Decision-Making, 71
   Investing in Regional Centers, 73
   Investing in Networks, 74
   Considering All Aspects of Governance, 76
   Components of Effective Governance, 77
   Leadership Development, 81
   Increasing Capacity for Enforcement and Monitoring, 82
   Findings and Recommendations, 83

6 Building Capacity in Ocean and Coastal Governance   89
   The Relationship between Science and Governance, 90
   Governance Mechanisms and Building Capacity, 90
   Assessing Governance Capacity, 92
   Instilling the Tools, Knowledge, Skills, and Attitudes Required to Practice
     Ecosystem-Based Management, 93
   Codifying Good Practices and Developing Certification Standards, 100
   Building a Culture of Learning and Self-Assessment: The Basis of Adaptive
     Management, 102
   Findings and Recommendations, 104

7 The Path Ahead: Strategic and Long-Term Approaches to Capacity-Building 106
   A Vision for the Future, 106
   Recommendations, 108

References   113

Appendixes
A Committee and Staff Biographies   123
B Panamá Conference 2006: Are We Meeting the Challenges of
   Capacity-Building for Managing Oceans and Coasts?   129
C Major Changes in Capacity-Building Since 1969   133
D Acronyms   140

Summary

The rapid decline of many ocean and coastal ecosystems and the global implications of current trends in oceans and along coasts have captured the attention of governments, national and international organizations, and many private organizations. A wide variety of projects, programs, and international conventions include capacity-building to strengthen the effectiveness of ocean and coastal governance and increase awareness of the benefits of maintaining ecosystem goods and services. Capacity-building, however, is not usually the primary focus. Capacity-building efforts typically are fragmented, lack standards for monitoring and evaluation, and are planned for too short a period to achieve and sustain effective ocean and coastal management.

WHAT IS CAPACITY-BUILDING?

Capacity-building describes programs designed to strengthen the knowledge, abilities, relationships, and values that enable organizations, groups, and individuals to reach their goals for sustainable use of ocean and coastal resources. It includes strengthening the institutions, processes, systems, and rules that influence collective and individual behavior and performance in all related endeavors. Capacity-building also enhances people's ability to make informed choices and fosters their willingness to play new developmental roles and adapt to new challenges. Capacity is about more than potential; it harnesses potential through robust programs to make progress in addressing societal needs and is fundamental to fostering environmental stewardship and improving the management of ocean and coastal areas and resources.

WHY IS THE NEED TO BUILD CAPACITY URGENT?

Globally, ocean and coastal ecosystems are under great stress because of the effects of human settlements and activities. Nearly 40% of the world's population lives within 100 km of coasts, and this proportion is expected to increase to 50% by 2015. The coasts and the oceans yield tremendous benefits to society through the extraordinary productivity of many marine habitats and the strategic benefits of a coastal location to trade, defense, industry, and food production. More than a billion people rely on the oceans and coasts as their primary source of food protein, and ocean and coastal environments support the livelihoods, economies, and quality of life of many communities.

Many alterations of ocean and coastal ecosystems have brought substantial benefits to society and improved the lives of billions of people. However, the density of development and intensity of resource use have also had serious impacts, including the conversion of habitat; modification of flows of water, sediments, and pollutants to the sea; changes in biological diversity; climate change; and drainage of excess nutrients into coastal waters. Those changes have in many instances weakened the ability of ecosystems to generate current and future ecosystem services. Continuing alterations affect ecosystem functioning and often impair the delivery of valuable services, such as nutrient recycling. Some 60% of global ecosystem services are degraded, and only a few services are increasing. When capacity to promote stewardship and practice ecosystem-based management is insufficient, the sustainability of ocean and coastal ecosystems is at risk.

CURRENT STATUS OF CAPACITY-BUILDING FOR OCEANS AND COASTS

This report examines the characteristics of existing capacity-building efforts by drawing on the expertise and observations of the members of the National Research Council's Committee on International Capacity-Building for the Protection and Sustainable Use of Oceans and Coasts and the presentations and discussions at a committee-led workshop in Panamá. The report assesses the strengths and weaknesses of past and current efforts to build the capacity required to improve the effectiveness and efficiency of ocean and coastal conservation and development. It recommends how such governance capacity can be strengthened. It discusses approaches to capacity-building that can bridge the gaps between planning and the implementation of effective management policies and plans of action. A major emphasis is placed on sustaining investments in capacity-building over the long term and on identifying how international partnerships can instigate and collaborate in long-term programs.

The study was funded by the Gordon and Betty Moore Foundation, the President's Circle of the National Academies, the David and Lucile Packard Foundation, the National

Oceanic and Atmospheric Administration, the National Science Foundation, the Marisla Foundation, and the Curtis and Edith Munson Foundation. The statement of task of the committee is shown in Box S.1.

Most capacity-building activities have been initiated to address particular issues, such as overfishing or coral-reef degradation, or to target a particular region or country in the developing world to address issues of sustainability and poverty. That pattern has resulted in ocean and coastal capacity-building activities that consist largely of fragmented, short-term training and education programs. There is little coordination among efforts that have similar goals or overlapping geographic coverage, and programs become isolated geographically and temporally. That fragmentation inhibits the sharing of information and experience, reduces opportunities to maintain new programs through consecutively funded efforts, produces gaps in governance, and makes it more difficult to design and implement management approaches that are consistent with the scale of the affected ecosystems. Fragmentation of efforts at multiple levels is an underrecognized barrier that needs to be overcome to improve capacity-building.

Box S.1
Statement of Task

The study will identify barriers to effective management of coastal and marine resources encountered in coastal nations, particularly in the developing world. The committee will examine current and past efforts to build the scientific, technological and institutional capacities required for countries to develop and implement effective coastal and marine resource policies. This review will include analysis of strategies for sustaining the benefits of capacity-building efforts over the long term. In carrying out its task, the committee will:

(1) Identify the types of information that would be required to form a foundation for policy decisions affecting the long-term health of coastal and marine ecosystems;

(2) Examine the roles of human resource development, establishment of appropriate institutions and infrastructure, and creation of a favorable policy environment in building legitimacy across a broad spectrum of society into oceans-related programs; and

(3) Identify measures to link investment in capacity-building to "on-the-ground" results, using such analytical tools as economic cost-benefit, environment and development indicators, and transboundary diagnostic analysis.

The committee will recommend ways in which the United States and partner organizations, including governments, international bodies, and stakeholders, can help strengthen the marine protection and management capacity of other countries. This will include recommendations on how capacity-building activities can be translated into sustainable environmental and economic programs.

Other barriers have limited the effectiveness of capacity-building programs: (1) lack of an adequate needs assessment before program design and implementation, (2) exclusion of targeted populations from decision-making, (3) poor management structures that can lead to mismanagement or corruption, (4) incomplete or inappropriate evaluation procedures, and (5) the paucity of long-term programmatic monetary support and the lack of a coordinated and strategic approach among donors.

THE GOVERNANCE DIMENSIONS OF ECOSYSTEM STEWARDSHIP

Effective and long-lasting ocean and coastal stewardship can occur only when a predictable, efficient, and accountable governance system is in place. Successfully executed governance initiatives establish dynamic processes that are maintained by the active and sustained involvement of the public and stakeholders who have an interest in the allocation of coastal resources and the mediation of conflicts. The processes of governance are expressed by three mechanisms: markets, governments, and the institutions and arrangements of civil society. How those mechanisms interact with one another is complex and dynamic and needs to be a focal point of future capacity-building.

Many tasks are necessary to develop and sustain ocean and coastal ecosystem governance initiatives, and they require expertise in a variety of disciplines. However, most professionals are trained in a single discipline and have little exposure to or experience in other fields. Growing capacity for the stewardship of ocean and coastal ecosystems requires an ability to integrate across diverse perspectives and disciplines. Analysis of the condition and dynamics of an ecosystem, the forces of change, and ecosystem resilience requires a broad knowledge base and the ability to integrate what is known into a framework that addresses problems, builds on opportunities, and takes into consideration the area's culture and traditions. Capacity-building programs therefore need to instill the tools, knowledge, skills, and attitudes that address:

- How ecosystems function and change.
- How the processes of governance can influence the trajectories of societal and ecosystem change.
- How strategies can be tailored to the history and culture of the place.
- How to assemble and manage interdisciplinary teams.

The report explains how that combination of capabilities could and should be imparted through future capacity-building investments.

FEATURES OF FUTURE CAPACITY-BUILDING

The success of future capacity-building programs will require education and training opportunities, effective governance structures, and sustained economic support. The committee has identified the following attributes that would increase the effectiveness and efficiency of future capacity-building programs:

- **Documentation of changes in capacity through assessments that use a consistent set of criteria.** Regular assessments will be needed to help programs to adapt to changing needs in long-term capacity-building efforts. Some common criteria will facilitate comparisons through time and across programs, but assessments will need to be tailored to fit the circumstances and characteristics of specific programs.
- **Funding of capacity-building through diverse sources and coordinated investments by local, regional, and international donors.** Building sustainable programs requires longer-term support than is typically provided by individual donors.
- **Support of dynamic and committed leaders, usually local, to develop a culture of stewardship and to work with the community to develop and implement a plan of action to sustain or improve ocean and coastal conditions.** Effective leaders also serve as mentors and role models that can motivate future leaders.
- **Development of the political will to address ocean and coastal management challenges.** Political will requires building a base of support for ocean and coastal stewardship through greater awareness of its long-term societal benefits. Public discussion of the costs and benefits of environmental sustainability—stimulated by the mass media, information campaigns, and educational programs—will heighten awareness of and build political will for necessary changes in the processes of planning and decision-making.
- **Establishment of continuing-education and certification programs to build the capabilities of practitioners.** This will enable current and future generations of professionals to adapt and apply the best practices to ocean and coastal management in diverse settings.
- **Networking of practitioners to increase communication and support ecosystem-based management along coastlines, in estuaries, and in adjoining large marine ecosystems and watersheds.** The networks will facilitate collection and integration of information and knowledge, new technologies, and Web-based data management systems in support of locally implemented, regionally effective, ecosystem-based management.

- **Collaboration among programs in neighboring countries through the founding of regional centers to encourage and support integrated ocean and coastal management.** The centers would link education, research, and extension to address issues of concern in the region and provide an issue-driven, problem-solving approach to capacity-building.

RECOMMENDATIONS

Seven critical actions are recommended to establish sustained capacity-building that can adapt to the changing conditions in ocean and coastal ecosystems.

RECOMMENDATION: Future investments in capacity-building should be anchored by periodic needs assessments used to develop regional action plans.

A recurring theme among experienced practitioners in capacity-building programs is the importance of anchoring capacity-building in thorough needs assessments. Such assessments should refer to a baseline of environmental, social, and economic conditions and analysis of the existing governance structure. Periodic assessments will be required every three to five years to update priorities to address changes in the ecosystem and responses to them.

Assessments will be required for each region because the maturity, capabilities, challenges, and traditions of governance differ from one place to another. For example, capacity-building priorities in Southeast Asia and the best strategies for meeting them will be quite different from those appropriate to East Africa or Central America. The credibility of the periodic needs assessments will depend on the participation and buy-in of the major investors in capacity-building in that region. Each assessment should be designed to attract high-level attention to high-priority issues. The findings should form the basis of regional action plans to guide investments in capacity and set realistic milestones and performance measures. Action plans should include concrete agreements on roles and responsibilities of donors, who provide financial support, and doers, who share their tools, knowledge, skills, and attitudes to strengthen capacity.

RECOMMENDATION: Capacity should be built to generate sustained funding for ocean and coastal governance.

Capacity is grown through the cumulative efforts of doers and donors to develop self-sustaining programs for knowledge-based ocean and coastal ecosystem-based management. Ecosystem-based management requires a long time to yield the greatest societal benefits and to adapt to the rapidly changing conditions of ocean and coastal ecosystems.

In developing countries, however, where ocean and coastal change and the loss of critically important goods and services is most rapid, the dominant mode of investment in ecosystem-based management is the 2- to 10-year project, and most initiatives are funded for 5 years or less. That applies to initiatives on scales ranging from community-based projects to the large marine ecosystem programs supported by organizations, such as the Global Environment Facility.[1] Many promising efforts wither and die when external funding from the donor community or development banks ends. There is an urgent need to build awareness of this problem so that future programs can be designed and implemented with strategies for sustained financing. Guidelines should be developed to provide practitioners with the knowledge and skills required to apply market-based mechanisms, such as user fees, regulatory fees, beneficiary-based taxes, and liability-based taxes.

RECOMMENDATION: Capacity-building programs should include programs specifically designed to develop, mentor, and reward leaders.

One of the most commonly cited reasons for failure and lack of progress in ocean and coastal governance initiatives is the lack of political will. One strategy for building project momentum and broadening support is to identify, develop, mentor, and reward leaders. Leaders are gifted communicators who play a central role in navigating the process of assembling support for a course of action. Leaders are not necessarily practitioners who have technical skills, but they may emerge from any of the various doer communities. The capabilities of leaders should be built through specific programs designed to enhance leadership skills. Investments in leadership will be most effective when they are associated with a regional network of programs that facilitate sharing of new information and ideas and build solidarity among people working to achieve common goals.

RECOMMENDATION: Networks should be developed to bring together those working in the same or similar ecosystems with comparable management or governance challenges to share information, pool resources, and learn from one another.

Networks are cost-effective and efficient mechanisms for maintaining and building capacity. They foster the creation of learning communities on the basis of trust and mutual respect. Well-structured networks help communities to envision the bigger picture and reduce members' sense of isolation by building solidarity and a common purpose with each other. Networks associated with periodic regional assessments of needs and progress

---

[1]The Global Environment Facility is used here only as an example of a large organization, not to single it out.

can encourage discourse on and critical examination of what works and does not work and can thereby promote implementation of successful practices.

Networks are enhanced by periodic personal contacts, but much can be accomplished through well-structured and adequately maintained Web-based systems. Information systems designed to support the creation, capture, and dissemination of knowledge and directed specifically at enhancing capacity in the practices of ecosystem-based management to overcome the "implementation gap" could provide practitioners with the material to analyze successes and failures, to identify and resolve specific technical and policy issues, to recognize opportunities for transboundary collaboration, and to gain access to public education materials and meeting summaries produced by participating programs.

RECOMMENDATION: Regional centers for ocean and coastal stewardship should be established as "primary nodes" for networks that will coordinate efforts to fulfill action plans. These centers will require a contingent of experience-based professionals and infrastructure to serve as a resource for the entire network.

Decentralized networks and centers that combine research and education with outreach and extension are most effective in fostering discussion and implementing new approaches in the surrounding communities. In the United States, the Land-Grant University System and the National Sea Grant College Program illustrate how a network of institutions can foster the adoption of new practices in agriculture, aquaculture, public health, and education when there is a long-term, sustained effort. The adaptation and application of the integrated education-research-extension model on national and regional scales as a primary strategy for developing capacity for ocean and coastal governance would offer a powerful alternative to the current pattern of investment in expensive short-term and disconnected "projects."

RECOMMENDATION: Progress in ocean and coastal governance should be documented and widely disseminated.

The effectiveness of future capacity-building programs could be enhanced by careful examination and analysis of traditions of governance, of prevailing societal and environmental conditions, and of how the context influences the structures of and strategies for environmental stewardship. Periodic needs assessments should be used to document and analyze the evolution of selected ecosystem-based management initiatives in each region. It will be particularly important to integrate the often rich but scattered information and experience on ecosystem change and governance initiatives in linked watersheds, estuaries, and large marine ecosystems. The analysis should document changes in societal and environmental benefits generated by the application of ecosystem-based management principles and practices. A common conceptual framework should be used to document and analyze ecosystem-based management initiatives in diverse cultural, geographic, and biophysical settings to inform future capacity-building efforts and potentially help to build political will for new initiatives.

Regional programs build on successes achieved on smaller spatial scales. Large marine ecosystems adjoin coastal zones; both are influenced by the rivers, wetlands, and estuaries that transmit the effects of land-based activities to the sea. Hence, goals for the stewardship of marine areas will require efforts beyond traditional sector-by-sector planning and decision-making and beyond experience with ocean and coastal protected areas. Each sector of governance, from local community-based management to national ocean policies and from inland to offshore areas, should be coordinated and efficient.

RECOMMENDATION: A high-level summit should be held on capacity-building for stewardship of oceans and coasts. This summit should be held to demonstrate political will, with commitments to end fragmentation, and to build action plans for capacity-building based on regional needs assessments that integrate with other programs that address ocean and coastal stewardship issues.

Strengthening and coalescing political will among institutional leaders in government, nongovernmental organizations, and industry will be required to overcome the problem of fragmentation through the critical actions identified above. Political will is required to establish programs that focus specifically on capacity. One factor that has limited past efforts is that capacity-building is usually treated as an ingredient of programmatic efforts on specific topics. It has been identified as an ingredient of plans for the Global Ocean Observing System and is addressed in Agenda 21 of the United Nations Conference on Environment and Development and in the Millennium Ecosystem Assessments. In each case, the identification of capacity-building as a critical ingredient is valid, but the uncoordinated calls for increased capacity on specific topics result in the fragmentation of capacity-building efforts that typically have a lower priority than the other aspects of a program. What is lacking is a high priority for capacity-building in a program in its own right.

The committee calls for a high-level summit on growing capacity for stewardship of oceans and coasts to demonstrate political will, to commit to ending fragmentation, and to build an agenda for capacity-building that cuts across other programs that address ocean and coastal stewardship issues. Many meetings have been held and continue to be held, but often they are not at a high enough level to demonstrate political will and do not dedicate their agendas to capacity-building and the need to reduce fragmentation of efforts.

The phrase *high-level summit* is used to emphasize the importance of a meeting at an appropriately influential level to demonstrate political will. Various types of meetings could serve that function. It might be a follow-up to the World Summit on Sustainable Development, it could build on the United Nations Open-Ended Informal Consultative Process on Oceans and the Law of the Sea, or other venues may be appropriate.

Key leaders with a regional stake in stewardship of oceans and coasts should form the core of the summit, and people in capacity-building communities (doers and donors) should be engaged. The summit should involve governments, nongovernmental organizations, intergovernmental organizations, academe, and the private sector.

CONCLUSION

Ending the fragmentation of current programs that seek to grow capacity for ocean and coastal management and to improve stewardship will require a new, broadly adopted framework for capacity-building programs that emphasizes cooperation, sustainability, and knowledge transfer within and among communities. The condition and sustainability of the vital resources and services of oceans and coasts that are valued by societies around the world depend on increasing the global capacity for good stewardship. Developing nations face a steeper challenge to develop more sustainable ecosystem-based management practices that can be met in part through capacity-building. But all nations share a responsibility to develop the capacity and institutions for more sustainable management of the oceans and coasts that connect nations and continents around the globe.

1
INTRODUCTION

Communities in both the developed world and the developing world rely on the productivity and diversity of ocean and coastal waters for many necessities and amenities. As populations increase, particularly along the coasts, those waters and the resources they contain become subject to ever greater alteration and exploitation. Climate change will increase the challenges of living and working in coastal areas because of higher sea level, warmer waters, changes in storm intensity, and shifts in the diversity and abundance of marine life. There is an urgent need to educate and advise policy-makers about these challenges so that strategies for adaptation can be developed. Marine ecosystems are important to human well-being, but there is a widespread lack of the tools, knowledge, skills, and attitudes necessary to manage the oceans and coasts to sustain an equitable suite of benefits for current and future generations. To prevent deterioration and potentially irreversible loss of valued marine resources, communities need to acquire the physical, human, and economic capital necessary to develop a scientific basis of management and to educate and inform decision-makers and citizens so that they become successful stewards of their environment.

Developing the capacity for ocean and coastal stewardship is a challenging undertaking. Ocean and coastal stewardship is complicated by many constraints, including (1) the challenges of anticipating and managing unprecedented changes in coastal ecosystems; (2) the difficulty and expense of monitoring marine ecosystems; (3) the legacy of outdated attitudes and knowledge about ocean and coastal ecosystems; (4) the fractured nature of ocean and coastal management; (5) the impacts of land-based activities on coastal resources and ecosystem services; (6) the insufficiency of means of documenting the effectiveness of governance systems; and (7) the relative scarcity of established mechanisms for planning and decision-making in the marine realm as opposed to the terrestrial realm, in which property rights and jurisdictional boundaries are more firmly established.

In addition, institutional systems are often complex, with overlapping jurisdictions, conflicting responsibilities, and legacies that make stewardship difficult. Those legacies frequently arose from the tradition of "freedom of the seas", a concept that governed international attitudes to ocean resources beyond the territorial seas and limited management to fragile treaties or voluntary agreements (Young, 1989). Even coastal seas have been treated as part of the public commons and have been subject to overexploitation because the tradition of open access, the dominance of short-term and narrow objectives, and the lack of alternative livelihoods often make it difficult to establish effective regulatory regimes.

Conflicts over access and tenure, piracy, illegal and unreported extraction and polluting activities, and tendencies toward overexploitation of shared resources and open-access resources all warrant improved governance. Resolving conflicting societal goals and values is difficult, and the existence of multiple interests and values means that success or failure of stewardship is largely in the eye of the beholder. Thus, capacity-building for stewardship of oceans and coasts in part involves establishing a process for decision-making that is viewed as legitimate by a broad spectrum of stakeholders. This report examines some of the key components of capacity-building programs for strengthening management and conservation of ocean and coastal resources for the sustainability of marine ecosystems worldwide.

ORIGIN OF THE STUDY

The Ocean Studies Board initiated this undertaking to examine how people and institutions in the United States could work in partnership with other governments, international bodies, and stakeholders to strengthen the ocean and coastal protection and management capacity of coastal nations. In addition, the study examines how capacity-building activities could be translated into sustainable environmental and economic programs. An ad hoc committee of international experts was assembled to review current and past efforts to build the scientific, technological, and institutional capacities and to identify barriers to effective management. The committee's statement of task is given in Box S.1. This study was funded by the Gordon and Betty Moore Foundation, the President's Circle of the National Academies, the David and Lucile Packard Foundation, the National Oceanic and Atmospheric Administration, the National Science Foundation, the Marisla Foundation, and the Curtis and Edith Munson Foundation.

The findings and recommendations of the committee are based on the shared experience of the committee members, discussions with representatives of the donor and academic communities, and selected literature on current efforts in building capacity for sustainable use of oceans and coasts. In addition, the committee benefited from presentations by scientists, engineers, policy-makers, regulators, nongovernmental organizations, and community leaders during its international workshop in Panamá (Appendix B); this workshop enabled the committee to receive input from the international community and to identify case studies that illustrate different approaches to capacity-building.

KEY CONCEPTS

Capacity-building: Capacity-building is "the sum of efforts needed to nurture, enhance, and utilize the skills and capabilities of people and institutions at all levels" (National Research Council, 2002)—locally, nationally, regionally, and internationally (National Round Table on the Environment and the Economy, 1998). Capacity-building increases knowledge, abilities, relationships, and values that enable organizations, groups, and individuals to strengthen the institutions, processes, systems, and rules that influence collective and individual behavior and performance in all endeavors. Capacity-building also enhances people's ability to make informed choices and fosters their willingness to play new developmental roles and adapt to new challenges (United Nations Environment Programme, 2004). Capacity is about more than potential; it harnesses potential through robust programs to make progress in addressing societal needs. Examples of capacity-building activities are described in Chapter 3.

Ecosystem-based management: Ecosystem-based management, or the ecosystem approach, applies current scientific understanding of ecosystem structure and processes to achieve more coordinated and effective management of society's multiple uses of and interests in the services provided by the ecosystem. Ecosystem-based management does not prescribe a particular outcome; instead, it acknowledges that changing the ecosystem can also change the services it provides. To sustain the ecosystem services that people want and need, an ecosystem-based approach incorporates diverse stakeholder perspectives and balances conflicting objectives to develop an integrated approach to management. The greatest depth of knowledge of human activities and ecosystem responses has thus far come from studies focused on particular sectors. Hence, ecosystem-based management incorporates past and current experience with sectoral management into the evolving analysis of ecosystem dynamics and incorporates new ventures into broader-based management.

The evolution toward ecosystem-based management is under way. Managers are beginning to make better use of existing knowledge, and adapting approaches based on experience with various management measures. As experience and the knowledge base increase and governance structures evolve to consider diverse interests in oceans and coasts and to foster discussion among sectors, management will become more attuned to ecosystem responses and limits. Those changes are taking place in developed and developing countries, and some developing countries are helping to lead the way.

Stewardship: Whether personal or institutional, *stewardship* commonly refers to a community[1] ethic adopted to ensure that natural resources are sustainably used and managed for maintaining quality of life of current and future generations. In the context of capacity-building for ocean and coastal management, the ethic of stewardship promotes sustainable use and conservation to maintain the health of ecosystems through knowledge-based systems of ocean and coastal management.

Governance: Governance is the societal structure, going beyond formal systems of government, that encompasses the values, mores, policies, laws, and institutions by which a society addresses a set of issues. It includes the fundamental goals, institutional processes, and structures that create the basis of planning and decision-making (Olsen, 2003). It encompasses the formal and informal arrangements, institutions, and values (Olsen et al., 2006a) that influence:

- How a resource or an environment is used.
- How problems and opportunities are evaluated and analyzed.
- What behavior is deemed acceptable or forbidden.
- What rules and sanctions are applied to direct how natural resources are allocated and used.

The processes of governance are expressed through the institutions and arrangements of markets, government, and civil society (Juda, 1999; Juda and Hennessey, 2001). Those entities interact with one another in complex and dynamic ways—from profit-seeking associated with markets to the laws, regulations, and taxation policies imposed by governments and the preferences and actions of civil society.

Doers and Donors: Doers and donors are the people and organizations involved in growing capacity for stewardship of oceans and coasts. Doers are the individuals or organizations that share their tools, knowledge, skills, and attitudes through direct engagement

---

[1]A community consists of a coalition of individuals that interact with each other and with other groups and individuals. A community may consist of families, neighborhoods, local advocacy groups, churches, nongovernmental organizations, and local governments.

with the people who stand to benefit from enhanced capacity. Donors, who may also be doers, contribute monetary support and physical assets to capacity-building.

Practitioners: Practitioners are the people who apply the knowledge gained from capacity-building efforts to accomplish the management goals of the community. They may also be involved in capacity-building efforts—in which case they would also be considered doers.

REPORT ORGANIZATION

This report is organized to review the challenges, constraints, and lessons learned from experience with capacity-building efforts. Throughout, the committee refers to particular projects and programs to illustrate points that are being made. It should be noted that the committee recognizes that there are many good examples that are not discussed in the report, so the inclusion of a particular program should be construed not as an endorsement but as an example that illustrates a particular aspect of capacity-building.

Chapter 2 describes the special challenges of achieving sustainable use of oceans and coasts. Chapter 3 analyzes the evolution and limitations of past and current capacity-building efforts to achieve that sustainable use and to promote stewardship of oceans and coasts. Chapter 4 identifies the barriers to and constraints on effective capacity-building. The remaining three chapters summarize the committee's view of the way forward in capacity-building for effective governance and stewardship. Chapter 5 focuses on aspects of capacity-building that are currently underemphasized aspects that have typically received less attention than the more science-based analysis of ecosystem change in existing capacity-building programs. Chapter 6 discusses ways to make capacity-building efforts more effective. Chapter 7 summarizes the committee's recommendations and presents a vision for future capacity-building efforts. Appendix A contains committee and staff biographies, Appendix B presents the agenda of the committee's workshop in Panamá, Appendix C summarizes the history of capacity-building, and Appendix D contains a list of acronyms used in the report.

2

The Challenges of Achieving Stewardship of Oceans and Coasts

HIGHLIGHTS

This chapter:

- Examines the challenges associated with achieving stewardship of oceans and coasts as the basis of discussions in later chapters on barriers to capacity-building and recommended strategies.
- Discusses the complex linkages between the natural resources of our oceans and coasts, the services provided by them, and human society's dependence on the sustainability of their healthy ecosystems.
- Details the current status of the ecosystems, the challenges facing them in the future, and the critical importance of effective ocean and coastal stewardship and ecosystem-based management to address the challenges.
- Reviews the capacities needed to address the challenges.

OCEAN AND COASTAL ECOSYSTEMS AND SERVICES

Nearly 40% of the world's population is concentrated in the 100-km-wide strip of coast along each continent, although it comprises only 5% of the habitable land area on Earth (Millennium Ecosystem Assessment, 2005a). Many of the coastal residents in developing and developed countries depend directly on ocean and coastal ecosystems for their livelihood. The extraordinary productivity of many ocean and coastal habitats and the strategic benefits of a coastal location to trade, defense, nonrenewable natural resources (such as oil, natural gas, and minerals), industry, and food production have made the ocean and its shoreline uniquely important. Inland residents also benefit from ocean and coastal resources and habitats through the provision of many services, such as the production of seafood and recreational opportunities and the buffering influence of the ocean on climate.

Seafood is one of the obvious benefits obtained from ocean ecosystems, but many other benefits, collectively called "ecosystem services," are derived from the ocean. *Ecosystem services* refers to the natural processes that provide renewable resources, maintain biodiversity, and sustain and fulfill human life (Millennium Ecosystem Assessment, 2005a). Ocean and coastal ecosystems provide many services through the diverse, interconnected assemblages of habitats that collectively provide such resources as food, water, building materials, and economic, recreational, educational, and inspirational opportunities. The Millennium Ecosystem Assessment (2005a) divides ecosystem services into four categories: provisioning, regulating, cultural, and supporting (Figure 2.1); the supporting services assist those in the other three categories.

Moreover, ecosystems are connected to and interact with adjacent ecosystems. The intimate connections across the land and the sea—transferring nutrients, energy, and material in both directions—are only now beginning to be quantified. Humans are an integral part of the ecosystems through their use of and impact on the environment and associated resources. Extraction of resources, agriculture, forestry, urbanization, aquaculture, port dredging, and waste disposal are some of the human actions that can change the

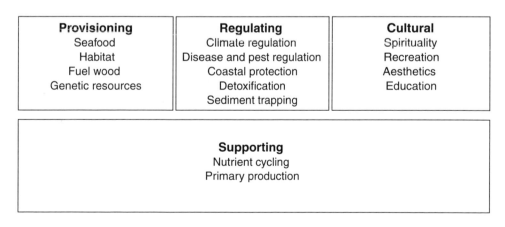

FIGURE 2.1 Categories of ecosystem services and examples in ocean and coastal ecosystems. Source: Modified from Millennium Ecosystem Assessment, 2005b; United Nations Environment Programme, 2006.

physical, chemical, and biologic nature of ocean and coastal ecosystems. The coupling of coastal development with lack of management or poor management of pollutants and living resources has caused the loss of many coral reefs, mangroves, marshes, beaches, and other coastal habitats and undermined the livelihood of millions of people worldwide. Once the coastal habitats are damaged or lost, restoration is difficult and expensive. In addition, the costs brought about by loss of services, such as coastal protection, are incurred for long periods (Moberg and Rönnbäck, 2003). Humans thus both depend on and modify ecosystems in ways that affect future delivery of services (Figure 2.2).

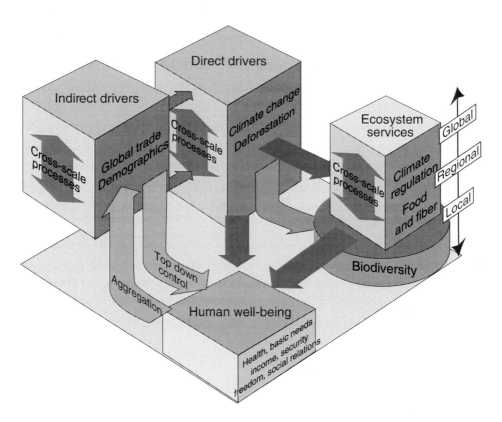

FIGURE 2.2 Relationship between human dependence and impacts on ecosystem services. Biodiversity, the diversity of life on Earth, provides the foundation for the delivery of services. The provision of ecosystem services is affected by indirect drivers, such as population and global trade, that in turn affect direct drivers of change, such as habitat conversion and climate change. Human health, prosperity, and well-being depend directly on ecosystem services that operate on a variety of spatial and temporal scales. Sources: Modified from Millennium Ecosystem Assessment, 2005b (reprinted with permission from World Resources Institute); Carpenter et al., 2006 (reprinted with permission from the American Association for the Advancement of Science).

CHALLENGES TO OUR OCEAN AND COASTAL ECOSYSTEMS

Current State

The rates, magnitude, and diversity of changes in ocean and coastal ecosystems have increased substantially—in many cases exponentially—over the last century (Vitousek et al., 1997; Lubchenco, 1998). The combination of higher global population, increased coastal population, and a more technologically dependent lifestyle has increased the pressure on natural resources and caused significant alterations to ocean and coastal ecosystems (Pew Oceans Commission, 2003; U.S. Commission on Ocean Policy, 2004; Millennium Ecosystem Assessment, 2005a; United Nations Environment Programme, 2006).

Widespread alterations include the conversion of habitat (such as mangroves to shrimp ponds or towns and cities, wetlands to agricultural fields, and sandy beaches to concrete seawalls), modification of flows of water and sediments to the sea, changes in biologic diversity (such as loss of genotypes, populations, and species and introduction of nonnative species), climate change (such as sea-level rise, acidification, higher ocean surface temperatures, and changes in storm intensity), and drainage of excess nutrients and pollutants into coastal waters.

Those changes often arose from activities that improved the lives of billions of people, but in many instances they also weakened the ability of ecosystems to generate current and future services. With continuing alteration, many valuable ecosystem services could be lost According to the Millennium Ecosystem Assessment, 60% of global ecosystem services are degraded, and only 4 of 24 services are increasing (Millennium Ecosystem Assessment, 2005b; United Nations Environment Programme, 2006; Figure 2.3). Those on the increase are primarily provisioning services.

The unintended consequence of a focus on provisioning services has been the erosion of regulating, cultural, and supporting services. For example, construction of shrimp ponds in a mangrove forest is intended to produce more food, but alteration of the mangrove ecosystems results in loss of the suite of services they provide. At least 30% of mangroves around the world have been lost because of a combination of coastal development and shrimp aquaculture (Valiela et al., 2001). Often, decisions are made to convert habitats, such as mangroves, without consideration or awareness of the trade-offs in ecosystem services and potential impacts on the communities benefiting from these services.

Seafood from wild stocks is a major provisioning service that is threatened, in many cases by unsustainable fishing practices, often combined with loss of fish habitats. According to the Food and Agriculture Organization of the United Nations (2007a), about one-fourth of the world's marine fisheries are overexploited or depleted. Although aqua-

	Enhanced	Degraded	Mixed
Provisioning Services	Crops Livestock Aquaculture	Capture fisheries Wild foods Wood fuel Genetic resources Biochemicals	Timber Fiber

Regulating Services	Carbon
	  sequestration
		Freshwater
		Air-quality
		  regulation
		Regional and local
		  climate regulation
		Erosion regulation
		Water purification
		Pest regulation
		Pollination
		Natural-hazard
		  regulation
			Water regulation
			Disease regulation

Cultural Services		Spirituality and
		  religion
		Aesthetics
			Recreation and
			  ecotourism

FIGURE 2.3 Ecosystem services balance sheet. Global status of 24 ecosystem services on which there is sufficient information to evaluate whether they are "enhanced" (increasing), "degraded" (decreasing), or "mixed" (increasing in some parts of the world and decreasing in others). Services are categorized as provisioning, regulating, or cultural. Information to classify supporting services is insufficient. Sources: Modified from Millennium Ecosystem Assessment, 2005b; United Nations Environment Programme, 2006.

culture has much promise for helping to provide food for future generations, the challenge is to ensure that it is conducted in a sustainable fashion. Salmon and shrimp aquaculture operations use wild-caught fish as a major component of feed, essentially converting fish of low economic value into more popular varieties. Substantial growth of other farmed carnivorous fishes (for example, cod) or of fish ranching (for example, tuna) is expected and will similarly require wild-caught fish for fishmeal and fish oil until fish feeds with reduced or no use of wild fish are developed to avoid unsustainable exploitation of wild stocks. In addition, aquaculture operations often pollute marine habitats, cause habitat degradation, and raise conflicts with other users of marine areas.

The downward trend in provision of ecosystem services is not inevitable, and degraded ecosystems may improve with better management or restoration efforts. Many changes in practices, policies, and governance will be needed to prevent further loss of marine ecosystem services.

Globalization

Globalization is driving change in unprecedented ways and is enabled by advances in communication and by increased mobility of people and commodities. Products that were previously exchanged locally are now bought and sold in global markets. For example, farmers in rural villages in India can now access the Internet to check the current price of a commodity and to negotiate an appropriate price with a buyer. Fish markets have become international, and the industrialized nations consume a disproportionate share, given that most of the fish are caught in the waters of developing nations.

Globalization presents new challenges to stewardship of oceans and coasts as changes in environmental conditions alter ecosystem processes and change the quantity, quality, and spatial and temporal distribution of ecosystem goods and services. Globalization also increases demand for some ecosystem services, for example, rising demand for fish products because of the globalization of markets and increases in tourism opportunities because air travel has become more available and affordable.

Globalization will increase the challenges for resource stewardship and will require an ability to manage despite uncertainty in the status and prospects of marine ecosystems. It will require institutional flexibility to adapt to changing conditions. Resource sustainability in one location will depend on the quality of resource stewardship in other areas of the ocean, so capacity is required at both the community level and the regional and global level.

Climate Change

Climate change adds a dimension to the challenges of sustaining ocean and coastal resources. The consequences of a warming climate include inundation caused by sea-level rise; acceleration of coastal erosion; changes in the intensity, distribution, and frequency of tropical storms; shifts in precipitation patterns; changes in the distribution and abundance of valuable marine species, including marine mammals, fish, and coral; and the frequency of coral bleaching and death (Intergovernmental Panel on Climate Change, 2007). In parallel with climate change, the ocean is becoming more acidic because of absorption of carbon dioxide from the atmosphere (The Royal Society, 2005). Ocean acidification has potential for widespread effects on marine ecosystems by inhibiting calcification, threatening the survival of coral-reef ecosystems, inhibiting the growth of calcareous algae at the base of the food web, and stunting the growth of calcified skeletons in many other marine organisms, including commercial fish species (Caldeira and Wickett, 2003; The Royal Society, 2005).

The ocean also provides regulating services such as carbon sequestration. However, an increase in ocean temperature due to climate change could change ocean circulation patterns and suppress the upwelling of nutrient-rich waters. The nutrients stimulate phytoplankton growth (sequestering carbon) and support highly productive marine ecosystems. Marine-plant photosynthesis fixes about 50 gigatons of carbon per year, roughly as much as is fixed by terrestrial plants, and thus accounts for a major fraction of organic carbon in the global carbon cycle. Hence, ocean acidification and higher surface water temperature could have dramatic effects on marine ecosystems and reduce the capacity of the ocean to moderate the climate and atmospheric carbon dioxide.

MOVING TOWARD OCEAN AND COASTAL STEWARDSHIP AND ECOSYSTEM-BASED MANAGEMENT

Human uses of oceans and coasts (such as fishing and other resource extractions, coastal development, and tourism) are imbedded in ecosystems and interact with natural processes to influence the complex dynamics of ecosystems. Indeed, humans are part of ocean and coastal ecosystems. The inescapable relationship between human activities and ecosystem dynamics has led to the widespread acknowledgment that management of the activities should be ecosystem-based. Ecosystem-based management is also known as the ecosystem approach. In 2002, the United Nations World Summit on Sustainable Development in Johannesburg, South Africa (United Nations, 2002), called for the "use of diverse approaches and tools, including the ecosystem approach."

Human well-being in the developing world and the developed world will depend on the ability of ocean and coastal ecosystems to provide a suite of ecosystem services. It will be a considerable challenge to make the transition to more sustainable practices and policies with the goal of providing services for the common good for both current and future generations. Ecosystem-based management, as characterized in Chapter 1, should be applied to meet the challenge. Improvements can be made in the short term by using knowledge better, acting cautiously to reduce the risk of undesirable outcomes, and including diverse societal goals and values in management decision-making. In the longer term, ecosystem-based management will benefit from research to increase ecological knowledge and implement more inclusive and responsive governance structures. Part of the challenge is to develop the human, institutional, and technological capacities for the evolution toward ecosystem-based management.

The application of ecosystem-based management is an evolutionary, as opposed to a revolutionary, process that can be implemented as the base of ecosystem knowledge increases. One example of the shift toward ecosystem-based management is described in a National Research Council (1999) report, *Sustaining Marine Fisheries*. The report addresses how fisheries management can move toward ecosystem-based management and concludes that a "significant overall reduction in fishing mortality is the most comprehensive and immediate ecosystem-based approach to rebuilding and sustaining fisheries and marine ecosystems." In the evolution from sectoral management to an ecosystem-based approach, available knowledge about ecosystems can be applied to account for effects outside the sector in recognition of and with respect for the values of people who are not participants in the sector.

The Role of Science in Ecosystem Stewardship and Governance

Knowledge that is relevant to good stewardship comes from many sources, including elders, cultural practices, communities, local resource users, nongovernmental organizations, the private sector, governmental agencies, and academia. Ocean and coastal sciences, including relevant social sciences and applied sciences concerning the management and governance of ocean and coastal resources, are evolving. Those disciplines require approaches that engage local stakeholders effectively. Capacity-building involves the exchange of information and expertise between the builders and the local people who seek assistance. A serious challenge in today's world is the connection of researchers generating knowledge with those who should be aware of or need to use that knowledge and translate it into action. Ideally, connections operate in both directions; users pose questions for researchers to investigate, and researchers share new knowledge with users. When researchers and users are the same people or when they live close to one another or have established communication channels, knowledge transfer can be accomplished more effectively and efficiently.

In considering the relationships between knowledge and action, it is important to clarify the role of science in decision-making. The committee concludes that one critical role of science is to inform individuals, institutions, and society during the decision-making process. Science should not dictate decisions but inform them in the following ways:

- Discover how natural, social, and coupled social-natural systems work.
- Document changes in these systems.
- Anticipate likely outcomes of the changes in view of an understanding of the workings of the systems.
- Develop and evaluate options for alternative trajectories.

New knowledge about ocean and coastal ecosystems is not always communicated to decision-makers effectively. Scientific information is often complex and nuanced and can contain uncertainties that are difficult to convey to a nontechnical audience. Moreover, many academic scientists have little understanding of the needs, culture, or language of different users, and this poses additional barriers to communication. If decision-makers are to be informed by science, they need to have access to scientific information that is understandable, relevant, credible, and useful (National Research Council, 2004).

Scientific evaluation is based on observations; sustained, long-term observations create the underpinning of scientific advice on stewardship of oceans and coasts. However, establishing and maintaining a global, integrated, and coordinated system of ocean observations to serve societies worldwide is a major undertaking—even components of the existing observing system are still far from complete. Much of the observing system operates through components of research programs and has no plan for sustained observations. Two-thirds of the world ocean is in the Southern Hemisphere, whereas most of the observing capability lies in the developed countries of the Northern Hemisphere (Partnership for Observation of the Global Oceans, 2001). Full implementation of a global observing system and full exploitation of the system for stewardship of the oceans and coasts will require participation from countries worldwide.

In many instances, similar types of ecosystems exist in geographically disparate parts of the globe (for example, coral reefs, mangroves, and coastal upwelling ecosystems). Knowledge about changes in an ecosystem type in one part of the world can shed light on likely changes elsewhere and thus improve management. For example, coastal upwelling ecosystems, which collectively represent 1% of the world's oceans but produce 20% of the world's fisheries, are found off the west coasts of Africa, Europe, North America, South America, and India. Many coastal upwelling systems appear to be undergoing substantial changes that are resulting in zones of low or no oxygen (hypoxic or anoxic zones). Knowledge of what happens in one or more of these systems can inform management of geographically distinct but ecologically similar ecosystems.

Developing New Strategies for Ecosystem Stewardship

Because humans are a part of the ecological system, human "systems" are intrinsically coupled to ecological systems. The coupled systems are inherently complex and characterized by nonlinear dynamics that occur over multiple spatial and temporal scales. Change can be abrupt, with little warning if a threshold is reached. Once a threshold has been crossed, a system may fail to return to its previous state even with the release or reversal of the pressures that caused the change. Coupled human-ecological systems are prime examples of complex adaptive systems (Levin and Lubchenco, submitted; Leslie and Kinzig, in review).

New fields of scientific study are emerging in response to the increasing impact of human societies on ecosystems, particularly over the last decade (Millennium Ecosystem Assessment, 2005c, d; Lotze et al., 2006). Called coupled human and natural systems (Liu et al., 2007, in press), science and technology for sustainability, and sustainability science (Kates et al., 2001), these fields are attempting to understand the linkages between natural and social systems, with a focus on emergent properties, knowledge, indicators, and tools for management and policy.

Resilience thinking is one new approach to addressing the decline in the capacity of communities, ecosystems, and landscapes to provide essential services. The intent is to recognize the complexity and variability of ecosystems, including the human component, and to build systems that can adapt to incorporate new knowledge or adjust to changing conditions. Management would shift away from production targets (such as tons of seafood caught or farmed) to management for resilience of the ecosystem (Walker and Salt, 2006).

Capacities Required for Effective Stewardship

Effective stewardship depends on capacities that are multidisciplinary—incorporating observations of the physical and chemical environment; ecosystem properties, processes, and human impacts; and biodiversity—so building capacity will entail such factors as human resource development through education and training, institutional and infrastructure development, and the creation of favorable policy environments that encompass a variety of public and private stakeholders. Strengthening governance skills in law, regulation, compliance, enforcement, monitoring, and evaluation is an important part of this mix and can help to ensure that advances are not short-lived.

Developing Required Expertise

Stewardship of oceans and coasts involves many sectors of society and depends on the expertise of many disciplines to address the complex challenges discussed earlier in this chapter. Capacity is required in the following disciplines:

- Natural sciences, such as mathematics, biology, chemistry, physics, geology, oceanography, and ecology.
- Social sciences, such as economics, sociology, anthropology, political science, geography, and law and government.

- Engineering—such as civil, industrial, mechanical, electrical, and chemical—and computer science and information technology.
- Business management, such as accounting, finance, and marketing.
- Professional skills, such as teaching, vessel operation, fishing, hatchery operation, and food processing.
- Organization management skills, such as human resource management, strategic leadership, conflict management and resolution, negotiation (alternative dispute resolution), strategic environmental planning, budget management, and project evaluation.

Capacity to Translate Knowledge into Action

Addressing the many complex facets of resource stewardship will require the capability to work across disciplines. There will be a need for scientists who understand management processes, managers who understand the strengths and limitations of science, and people who understand the role of institutions and legal instruments in governance for stewardship of oceans and coasts. Furthermore, stewardship requires the capacity to observe the environment, analyze data, and translate observations into information that is readily understood by the general public.

New programs or mechanisms to connect fundamental academic marine science and scientists to the public, policy-makers, nongovernmental organizations, the mass media, and the private sector are beginning to emerge. Some programs, such as the Aldo Leopold Leadership Program (Box 2.1), focus on training scientists to be better communicators. Others, such as the Partnership for Interdisciplinary Studies of Coastal Oceans, integrate outreach with research programs. Still others—such as the International Council for the Exploration of the Seas, the Communication Partnership for Science and the Sea, the National Center for Ecological Analysis and Synthesis, and the World Conservation Union with its various commissions, including the World Commission on Protected Areas, the Commission on Ecosystem Management, and the Species Survival Commission—produce consensus statements, syntheses, or assessments of knowledge on particular topics.

Ultimately, stewardship of oceans and coasts depends on a literate civil society. Literacy shapes societal values and public opinion and contributes to more informed decision-making. It can engender the political will to resist pressure from special interests in favor of decisions for the public good. It can also lead to public support for management, including compliance with regulations and intolerance for violators; for volunteers that participate in stewardship activities; and for monetary contributions to pay for capacity-building and stewardship activities.

Box 2.1
The Aldo Leopold Leadership Program

The Aldo Leopold Leadership Program (2006) advances environmental decision-making by providing academic scientists with the skills and connections needed to be effective leaders and communicators. Each year, up to 20 academic environmental scientists in North America are selected to receive intensive and analytic experiential training, expert consultation, and peer networking. During a two-week intensive training program, Leopold Leadership Fellows hone skills in communicating the science associated with complex environmental issues to the mass media, policy-makers, business leaders, and other nonscientists. More than 100 past fellows are actively engaged in scientific outreach in issues ranging from marine conservation science and river restoration ecology to effects of global climate change on human health. The target audience for the program is mid-career academic environmental scientists. Academic scientists are seen as trusted scientific voices, but they typically have no training in or understanding of effective ways to share their knowledge with nonscientists.

FINDINGS AND RECOMMENDATIONS

Ocean and coastal ecosystems are inextricably linked with humans and human well-being. Pressures on ocean and coastal ecosystems are likely to increase substantially as the human population continues to grow, as more people move to coastal areas, and as dependence on ocean and coastal resources increases. Improved understanding of the gains and losses in using and modifying ecosystems is urgently needed so that the trade-offs of various management options can be taken into account by decision-makers. In addition, alternative management approaches need to be developed and to coalesce under the concept of ecosystem-based management.

Building the capacity for implementing ecosystem-based management will, for example, require cultural literacy, interdisciplinary approaches, monitoring capabilities, communication skills, and channels to adapt concepts to a local or regional context. Future capacity-building in support of ocean and coastal stewardship should recognize the combination of tools, knowledge, skills, and attitudes that will be required to address the many dimensions of complex, ever-changing ecosystems efficiently. Capacity-building will require the shared expertise of both interdisciplinary teams and individuals.

The ocean connects the various parts of the globe physically through the currents that move water between the poles and the equator and between the surface and the deep and economically through global shipping routes and international markets for marine goods and services. More integrated approaches to stewardship—such as approaches that cross multiple sectors, focus on long-term benefits, adopt an ecosystem approach, and take full advantage of new knowledge—will be necessary to advance a global culture of stewardship.

Modern information and communication training and technology should be included in the creation of capacity-building programs to share expertise and lessons learned locally, regionally, and globally. New methods will be needed to translate scientific results into useful, relevant, and accessible information for decision-making. Both doers and donors should look for ways to apply new technologies and knowledge and improved communication methods.

3

GROWING CAPACITY FOR STEWARDSHIP OF OCEANS AND COASTS: A WORK IN PROGRESS

HIGHLIGHTS

This chapter:

- Details the people and organizations engaged in growing capacity for stewardship.
- Provides examples of capacity-building efforts on various scales of organization.
- Notes that current activities often lack coherence in providing sustained funding, recognizing ecosystem boundaries (as opposed to political boundaries), and establishing linkages among the various levels of government.

A number of approaches can be used to implement capacity-building programs, including understanding the players and their roles in the process; developing formal and informal education, research, communication, and training programs; reaching out to communities; and networking. Through such efforts, infrastructure that can ensure that communities have the tools, knowledge, skills, and attitudes necessary to become effective stewards of their ocean and coastal environments can be developed. Because knowledge evolves, environments change, and new stewards come of age, flexibility and adaptability are important attributes of capacity-building processes and institutions. Capacity-building as an activity has evolved over the last 40 years; Appendix C presents a brief history.

INVESTORS AND INVESTMENTS IN CAPACITY-BUILDING

Identifying the Investors: Doers, Donors, and Practitioners

People and organizations involved in growing capacity for stewardship of oceans and coasts may be involved as "doers" or "donors." Doers are the people or organizations that share their tools, knowledge, skills, and attitudes through direct engagement with those who stand to benefit from enhanced capacity. Donors, who may also be doers, contribute monetary support and physical assets to capacity-building.

A third category is the "practitioners" of ecosystem-based management (see Chapter 6), the primary targets of capacity-building. They work to steer the processes of governance and management toward stewardship outcomes. The practitioners are a subset of the many social and natural scientists, government officials, representatives of various stakeholder groups, and others that become engaged in an ocean or coastal management effort. They are usually directly engaged in the design and day-to-day administration of ocean and coastal management initiatives and in the management of interdisciplinary teams of specialists who are working to understand the issues involved in the management of an ocean or coastal ecosystem.

Institutions that address the use and health of the marine environment vary in size and scope, ranging from small teams of local villagers to informal assemblages of concerned citizens to large bureaucracies and intergovernmental organizations and other international bodies. Similarly, the scale of investment in growing the capacity of those institutions to address ocean and coastal stewardship varies from projects at small sites to multiyear programs that include many sites and sometimes many countries.

The major donors in ocean and coastal capacity-building typically are government agencies in both the developed world and the developing world. Most developed countries support capacity-building in ocean and coastal governance within their own institutions and through a variety of foreign-assistance programs. Substantial investments are made by the United States, the Nordic countries, the European Union, Canada, Japan, Taiwan, the People's Republic of China, and Australia. Many charitable foundations have supported capacity-building efforts in ocean-related matters, such as the Gordon and Betty Moore Foundation, the David and Lucile Packard Foundation, the Pew Charitable Trusts, the Ocean Foundation, the Sloan Foundation, and the Nippon Foundation.

Governments of developing nations may also be donors. They often play a central role in supporting capacity-building through programs to increase expertise in the science and management of marine and coastal resources. Some developing countries contribute to building capacity in other countries, particularly within their own regions, through partnerships, such as the Sustainable Fisheries Livelihoods Programme in West Africa.

Their investments may be augmented or even overtaken by the short-term, usually large, investments of international development banks, diverse multilateral and bilateral donors, and nongovernmental organizations (NGOs). However, governments have the primary responsibility for ensuring that investments in capacity-building yield results in the form of improved management of ocean and coastal resources; this requires a commitment to maintain and support the human and institutional resources grown by capacity-building programs.

Multilateral institutions that support capacity-building include the World Bank, the Global Environment Facility, the Inter-American Development Bank, the Asian Development Bank, and the United Nations and its specialized bodies, such as the Food and Agriculture Organization. NGOs in developed and developing nations contribute to education and outreach, surveillance and enforcement, scientific research and monitoring, management of visitor use in protected areas, and other management activities. Large environmental NGOs—such as The Nature Conservancy, the Wildlife Conservation Society, the World Wildlife Fund, the World Conservation Union (IUCN), and Conservation International—invest in capacity-building for ocean and coastal stewardship; and foreign-assistance agencies, such as the U.S. Agency for International Development and the UK Department for International Development, have established programs with NGOs to strengthen their structure and their ability to serve communities.

Direct and indirect private-sector investments in coastal stewardship and public private collaborations are increasing (Glasbergen, 1998; Hudson Institute, 2006). They include developer-financed conservation, restoration, and rehabilitation projects to comply with such regulations as policies for "no net loss" that require mitigation when wetlands are drained for development. There are also public-private partnerships, such as the teaming of municipal governments with chambers of commerce on watershed initiatives, and private financing of public-sector resource management, such as the generation of conservation funds through licensing fees (for example, for fishing, hunting, diving, and other recreational activities).

The diversity of investment mechanisms and sources, spanning local community efforts and national programs, illustrates the many approaches to funding capacity-building. As discussed below, there is no central "clearinghouse" that catalogs these activities, so there is little or no coordination of efforts, and funding often falls short of creating a stable foundation for sustaining programs.

The Need to Know and Monitor How Much Is Invested

It is nearly impossible to distinguish program funds dedicated *specifically* to ocean-related capacity-building. The committee believes that that is because of the following factors:

- Absence of a unified and accepted definition of *capacity-building*, which is often used as an umbrella term for development.
- Unclear definition of the components of capacity-building, especially distinctions between human and institutional components, at the project and program level.
- Lack of standardized methods for reporting data on ocean and coastal sectors.
- Lack of effective procedures for tracking where money is spent.
- Inclusion of ocean and coastal governance support in the broad category of funding for environmental programs.

Although the exact amounts spent specifically on capacity-building are difficult to ascertain, the total investment in programs that contribute to various degrees in ocean and coastal stewardship is large. For example, a recent analysis of international funding for the management of large marine ecosystems (Olsen et al., 2006a) indicates that in 1997 the countries of the Organisation for Economic Co-operation and Development (OECD)[1] spent an estimated US$2.24 billion on fisheries management (Wallis and Flaaten, 2000), an amount equivalent to 6% of the value of OECD fisheries landings. Emerton et al. (2006) summarize the latest available data on the amounts and sources of global funding for all protected area management; the funding totals US$6.5 billion a year. For comparison, Spergel and Moye (2004) estimate that the operation of a global network of marine protected areas alone could cost between US$7-19 billion a year. Olsen and Nickerson (2003) report that the Chesapeake Bay Program, an example of adaptive ecosystem management on the scale of a large estuary and its watershed, spends about US$70 million per year on efforts that are linked directly to program goals.[2] Programs for the Great Barrier Reef of Australia and the Wadden Sea of the Netherlands, Germany, and Denmark each spend about US$20 million per year on management efforts (Olsen and Nickerson,

---

[1]The 30 member countries of OECD are Australia, Austria, Belgium, Canada, the Czech Republic, Denmark, Finland, France, Germany, Greece, Hungary, Iceland, Ireland, Italy, Japan, Korea, Luxembourg, México, the Netherlands, New Zealand, Norway, Poland, Portugal, the Slovak Republic, Spain, Sweden, Switzerland, Turkey, the United Kingdom, and the United States.

[2]The area of the Chesapeake Bay is 6,475 km^2, and its watershed extends over 172,000 km^2. Management of the bay involves primarily efforts to control and reduce nutrients and pollutants that flow into the bay and its tributaries and to restore riparian and aquatic habitat to sustain estuarine fisheries (Olsen and Nickerson, 2003).

2003).[3] Those numbers provide some sense of the scale of investments in ocean and coastal management, of which capacity-building is a strategic component.

HOW TO GROW CAPACITY

Capacity-building efforts are designed to address particular ocean and coastal problems given the size and capabilities of the institutions involved (local, regional, and global). Some capacity-building efforts have targeted particular geographic regions; others have been aimed at specific sectors or problems or are more generally focused on enhancing capability in academic or professional disciplines. Some capacity-building emphasizes technical training and education, or technical assistance, whereas other efforts concentrate on improving governance institutions and empowering existing or new groups to use their own resources and skills more effectively.

The ultimate goal of developing capacity for stewardship is to establish the institutions and cadre of professionals that will enable society to use and conserve ocean and coastal resources knowledgeably, taking into consideration the broad interests of society for current and future generations. That goal requires that laypeople trust and respect the advice and services of the institutions and professionals that support stewardship. The public entrusts professionals not only on the basis of their education and training but also through standardized systems for certifying or licensing them to provide evidence of their competence to provide the desired services (see discussion on professional standards in Chapter 5).

Education, Training, and Outreach

The primary modes of capacity-building for education, training, and outreach have been (1) academic degree programs with a focus on a specific discipline (such as biological oceanography, fisheries, economics, and political science); (2) academic degree programs, commonly at the master's level, that provide a broad foundation in both the natural and the social sciences as applied to management of natural resources, typically including the public-policy dimension; (3) narrowly focused short-term training programs tailored to the performance of a specific set of stewardship activities; and (4) outreach and extension programs that provide technical assistance.

Education, training, and outreach opportunities for community-based practitioners, whether within or outside a community, should take into account local circumstances,

---

[3]Australia's Great Barrier Reef covers an area of 347,800 km^2; and the Wadden Sea, an estuary bordered by the Netherlands, Germany, and Denmark, covers an area of 13,500 km^2.

including culture and mores. Ideally, programs take place within a community setting and provide an authentic environment for learning in both formal (classroom) and nonformal (for example, on-the-job) settings. On-the-job training allows local practitioners to learn the content in the specific context of the location where the tasks will be performed.

Participants in education, training, and outreach need connections to the target community and the basic tools, knowledge, skills, and attitudes to build upon. Local, regional, and national governments and multilateral organizations can provide opportunities for people to study at foreign institutions and acquire the capabilities necessary to support community efforts.

Academic Degree Programs

Academic degree programs designed specifically to prepare for careers in ocean and coastal management have advanced greatly over the last 30 years. They cover a wide array of curricula, pedagogic techniques, and institutional settings that deliver capacity-building over a wide range of age and education levels, including:

- School programs (typically K–12, with awareness and exposure promoting "environmental literacy").
- University programs focused on a specific social science or natural science.
- University master's-level programs that provide a broader foundation in natural science, social science, and public policy.
- University outreach and extension programs (such as elements of the U.S. National Sea Grant College Program and the outreach programs of universities in the Philippines).

Until recently, programs in ocean and coastal governance and stewardship were found only in developed nations. Increasing numbers of universities in developing countries now provide the necessary curricula and have the ability to tailor programs to the issues and needs of their nation or region.

Several academic programs provide interdisciplinary studies in coastal management and marine policy (Table 3.1). They include longer-term degree-granting programs that educate people in key disciplines and feature interdisciplinary skills. Specialized university-based centers provide technical assistance, training, and evaluation services to many ocean and coastal governance initiatives worldwide.

One particular challenge for all these programs is to remain current and avoid teaching antiquated approaches or perspectives. For example, ecosystem-based approaches are now favored for developing management policies in response to reports on the status

TABLE 3.1 Academic and Training Programs in Interdisciplinary Coastal Management and Marine Policy[a]

Country	Program
Australia	Institute for Coastal Resource Management (University of Technology, Sydney)
Canada	International Ocean Institute (Dalhousie University)
Fiji	Ocean Resource Management Program (University of the South Pacific)
Greece	Rhodes Academy of Ocean Law
India	International Ocean Institute (Indian Institute of Technology)
Japan	International Ocean Institute
Malaysia	Malaysian Institute of Marine Affairs
Malta	International Ocean Institute
The Netherlands	Netherlands Institute for the Law of the Sea
Singapore	Asia-Pacific Centre for Environmental Law (National University of Singapore)
Thailand	South East Programme in Ocean Law, Policy, and Management (Sukhothai Thammathirat University)
United States	Gerard J. Mangone Center for Marine Policy (University of Delaware)
United Kingdom	University of Wales, College of Cardiff

[a]This table is not exhaustive but lists examples to demonstrate the global distribution of existing programs in developed and developing countries.

of global ocean ecosystems, for example, as described by the Millennium Ecosystem Assessment (2005a, b, c, d) and the United Nations Environment Programme (2006). Although some sector-based programs (such as those in fisheries and aquaculture) have broadened and modernized to incorporate ecosystem perspectives, many programs still focus on maximizing production without emphasizing a sustainable balance between production and ecosystem impacts. Ideally, capacity-building increases the ability of practitioners to stay abreast of new information and applications to practices and policies.

Training Programs

Short-course training in ocean and coastal management ranges from an introduction to the concepts and tools of integrated forms of ocean and coastal management to highly specialized courses on single topics. For example, these short-course training programs can include:

- Specialized training programs delivered by a wide array of for-profit organizations (for example, in leadership training), international organizations (such as the Intergovernmental Oceanographic Commission [IOC], the International Ocean Institute [IOI]), universities, and NGOs.
- Training via networks that foster peer-to-peer relationships, the sharing of experience, and collaborative learning.
- Web-based systems, which are increasingly valuable for making information readily available, that can organize and sort knowledge on specific topics.

Some training programs are designed to be broadly applicable to a region or country; others are tailored to the geopolitical and cultural circumstances of a particular place. For example, the United Nations Train-Sea-Coast Programme (United Nations, 2006) has a basic curriculum in ocean and coastal management training that can be adapted to the needs of specific regions or countries. The program is based on a "learning by doing" pedagogic scheme that provides learners with practical experience while they learn broader issues and ways of doing things that connect directly to the targeted community. Tailoring capacity-building to a specific locale, as do the many programs devoted to community-based resource management, requires ascertainment of local needs and then development of ways to assist in capacity-building.

Training often targets issues that are specific to management situations often found in developing nations. For example, fisheries management often has to address challenges posed by a shortage of information on fisheries stocks and by ineffective management institutions. Similarly, the surge in aquaculture activities in coastal areas, especially in Asia, has increased demand for short-term training in aquaculture that usually focuses on technological challenges—such as production efficiency, product safety, and environmental impact—and institutional and socioeconomic challenges. Such broad topics as pollution, habitat degradation, transportation, and tourism affect most coastal areas and are commonly covered in management training. As with any program, however, the challenge is to ensure that the content and perspectives of the training are both current and appropriate to the place.

Some capacity-building efforts focus on training in highly specialized disciplines to achieve program objectives. For example, the IOC conducts training on harmful algal blooms, data and information management, and rapid assessment of marine pollution (Intergovernmental Oceanographic Commission, 2002), and there is increasing recognition of the need for training in social sciences, ecosystem-based management, and business to accompany more common training programs in natural sciences and engineering.

Responding to the inadequacy of technical capacity to address the design and maintenance of coastal infrastructure in the Caribbean region, the University of the West Indies in Trinidad and Tobago developed the following programs:

- Professional development and long-term capacity-building in coastal zone engineering and management.
- Long-term training in the design, construction, and maintenance of coastal infrastructure in the joint U.S. Agency for International Development and Organization of American States Hurricane Lenny Recovery Project (Charles and Vermeiren, 2002).
- Distance training through the introduction of postgraduate programs in distance-education packages, such as coastal-zone engineering and management courses at the undergraduate and graduate levels.

The University of Rhode Island Coastal Resources Center (CRC) and the United Nations Train-Sea-Coast Programme are models of management training with an emphasis on the tools, knowledge, skills, and attitudes that are needed to help people and organizations to address multiple uses of and threats to coastal resources in a specific locale. They use learning-by-doing methods to enhance managers' abilities to address local environmental issues.

Stewardship of offshore ocean areas requires training in international law and policy. Subjects include the geography of marine jurisdictions, treaty law, and conflict resolution and negotiation. Many environmental-law curricula touch on those marine subjects, but thorough treatment depends on specialized degree programs, such as are found in institutes or schools of marine affairs. A handful of institutes provide training in or information on the legal aspects of international ocean management. For instance, IOI, which is based in Malta, has training programs in international maritime law and law of the sea and has gone from providing major inputs into the early development of the United Nations Convention on the Law of the Sea to providing training programs for coastal communities. The IOI network promotes sustainable use of ocean space and resources through increased awareness, education, information distribution, and research and community initiatives. IOI has thousands of training-program alumni around the world, many of them in influential decision-making positions in their home countries or the United Nations (International Ocean Institute, 2005).

Outreach and Extension Programs

Sustained outreach and extension programs have proved highly effective in changing the practices of target groups in such diverse fields as agriculture, fisheries and aquaculture, and public health. The U.S. National Sea Grant College Program is one model for connecting colleges and universities with the community through extension; it is based on the larger U.S. Land-Grant University System model of cooperative extension (National Research Council, 1995). Several countries have adapted the U.S. National Sea Grant College Program model to their own needs or are exploring importing it (Wilburn et al., 2007). Mechanisms are needed to ensure that the content and usefulness of these programs are periodically evaluated and updated.

In addition to formal extension programs, ad hoc outreach programs can be successful in facilitating local conservation initiatives. In the Philippines, Silliman University initiated outreach to communities on Apo Island to improve stewardship of the coral reefs. Biology faculty used informal slide presentations to the community on Apo Island to initiate discussion of the ecology of coral reefs, the resource degradation caused by dynamite fishing and other destructive fishing practices, and benefits of conservation initiatives (Alcala, 2001). The Marine Conservation Education Program was instrumental in the establishment of a reef sanctuary supported by a written agreement between the local municipality and Silliman University (Raymundo, 2002). This example illustrates that properly executed outreach can foster discussion between the people who benefit from technical assistance and the research and educational community in universities and government institutions. Extension programs are most effective when they are linked to a supportive institution that has complementary capabilities and programs in research and education.

Outreach programs that began as short-term projects in many developing countries have often faltered when the projects have ended and the teams have been disbanded. A more effective strategy is to invest in building and sustaining permanent centers with diverse capabilities in each region. Permanent centers can respond to research needs on topics of direct relevance to management and can develop curricula that target regional issues.

The Brain-Drain Effect

Education and training opportunities are critical for the long-term sustainability of capacity-building programs. However, such efforts are wasted if those who are trained do not have the tools necessary to use their training, do not have jobs available to sustain their own livelihoods, or are lured away from their home communities by the possibility of increased wages elsewhere, as the committee learned from presentations and discussions at the workshop in Panamá (Appendix B). (For a general discussion of the scope and consequences of brain drain, see Lowell et al. [2004] and National Research Council [2005].) It is incumbent on capacity-building efforts to recognize the potential for brain drain and formulate methods to minimize its impact. The methods can include education and training to produce new personnel to replace those who leave, using the trained personnel themselves as trainers, linking scholarships and sponsored training abroad to obligations to return and work in the trainees' native countries, and ensuring that the social infrastructure—the ability to make a living with the skills acquired—is established so that there are incentives for trainees to return to their communities. In addition, retention may be increased by enabling scientists from the developing world to attend professional meetings and network among their peers in other nations to reduce a sense of isolation in their home countries (Third World Academy of Sciences, 2004).

Technology and Tools

Capacity-building includes the transfer of innovative tools and technologies to address ocean and coastal issues and to aid government decision-makers in fulfilling their ocean and coastal stewardship mandates. Financial support is required to provide appropriate training in the use of the tools and technologies and development of the capacity to maintain and update the data generated. That shortcoming has limited the success of some programs, as noted in an evaluation of World Bank support for capacity-building in Africa (The World Bank, 2005):

> [Technical assistance] has been effective when used for discrete and well-defined technical tasks and in the context of a clear [technical assistance] strategy that includes a phase-out plan. A majority of the projects reviewed support training individual staff, and projects have almost always achieved the targeted numbers to be trained. But public agency staff is often trained for specific tasks before they are positioned to use the training or before measures are taken to help retain them.

In designing a technology scheme, it is necessary to consider both the physical infrastructure needs (such as electricity, computers, satellite dishes, and printers) and the companion technologies, such as software in the appropriate language and user-friendly format. Without a holistic perspective and efforts to translate new technologies into tools appropriate to decision-makers, doers, practitioners, and communities will be unable to realize the highest potential of these facilities.

The development of information technology has revolutionized information transfer and accessibility. Individuals and communities can access and make available documents and information from remote locations, communicate globally in real time with audio and video, and exchange datasets, photographs, videos, and documents. By enabling the formation of virtual social networks, these communication tools have facilitated the exchange of information among communities from the local to the global scale. Moreover, they have created communities around particular issues. Thus, a small isolated community can be connected to a global infrastructure and become part of the global community.

Virtual networks constitute a venue for sharing and exchanging information, facilitate matching of user requirements with donor resources, and serve as a tool for linking activities that might otherwise be fragmented. Networks can also be used to identify and document shared problems on a given topic and generate an information base to share best practices, allow for standardization and harmonization of methods, and link training activities to infrastructure-building activities. (For more discussion of networks, see Chapter 5.)

Specific data management technologies, such as a geographic information system (GIS), serve as powerful decision-support tools for ocean and coastal resource management. GIS can be used to capture, store, integrate, and display geographic reference information (U.S. Geological Survey, 2007). Other technologies include the global positioning system (GPS), a global high-accuracy, satellite-based system used in radio navigation systems and other location-detection systems (Reece, 2000). A vessel monitoring system (VMS) uses onboard equipment with built-in GPS that sends a message at predetermined intervals. It reports a vessel's location and identification to independent observers at a central monitoring center. VMS has direct applications for fisheries, marine protected area enforcement, and maritime safety (Food and Agriculture Organization of the United Nations, 2007b).

One of the most ambitious technology-transfer efforts for capacity-building is associated with the Global Earth Observing System of Systems (GEOSS). GEOSS is envisioned as a system that starts with observations and uses them to provide information that will benefit such fields as health, climate, weather, ecosystems, biodiversity, disasters, and energy. The ocean and coastal observing components are included as the Global Ocean Observing System, an element of GEOSS that is being developed under the auspices of the IOC and the World Meteorological Organization. The design of those systems needs to extend from the ocean and coastal waters to shorelines and the associated watersheds; they are all critical links for developing effective ecosystem-based management and informed decision-making.

Strengthening Institutions

Enhancing the capacities of institutions to deal with ocean and coastal issues has emerged as a major focus of the capacity-building efforts of multilateral development banks and some bilateral aid organizations, particularly since the United Nations Conference on Environment and Development in 1992. In part, that occurred with the proliferation of new integrated coastal management (ICM) or ocean area management efforts at national levels (as opposed to community pilot projects). It also stemmed from the observation that many investments in human-resources development and infrastructure met with little success because of institutional weaknesses that resulted in the inability to transfer investments effectively, efficiently, and equitably. Institutions may include local NGOs, civic groups, schools, and local, regional, or national government agencies or programs. Institutions provide organizational support for conservation and management initiatives and often enhance their legitimacy and impact. Some government organizations and NGOs have also begun to invest funds selectively in institutional capacity development. Box 3.1 highlights institution-building for ICM. The efforts in institution-building contribute to the larger goal of increasing capacity for ocean and coastal governance, discussed in more detail in chapters 5 and 6.

Local and Community Institutional Capacity-Building

A great deal of the investment in building institutions has been aimed at coastal communities. When successful, institution-building can increase the communities' sense of identity and pride, build relationships between individuals and their communities, strengthen local government, and improve the capacity for local service delivery and information flow. Those are important steps in establishing conditions that allow citizens to become good stewards of their ecosystems. The results can be rewarding, especially where the scale and dynamics of the resources are appropriate for those of the local governance system.

To reduce stresses on ecosystems, some capacity-building programs train people to earn a living in ways that do not exploit ocean or coastal resources. An example is the San Lorenzo project CEASPA (Centro de Estudios y Acción Social Panameño [the Panamanian Center for Research and Social Action]). CEASPA is a Panamanian NGO that works on very small-scale local projects, including that of Achiote, a small rural community that borders a tropical-forest region for which protection is being sought. Funding for the project—which involved small-scale efforts organized around ecotourism, bird-watching, local empowerment, and development of self-esteem—was only about US$110,000 in 2005–2006, but it depended heavily on volunteers (personal communication, Charlotte Elton).

Another well-known example is the Chilean program of creating management and exploitation areas for benthic resources (MEABRs). The program instituted comanagement of several coastal areas where nontransferable fishing rights were used to allocate shellfish grounds exclusively to artisanal communities. There are over 185 MEABRs in Chile, and

Box 3.1
The University of Rhode Island Coastal Resources Center's Approach to Building Institutional Capacity for Integrated Coastal Management

Institutional capacity-building can be developed by using demonstration projects to develop a constituency at national and local levels. For example, global experience in ICM demonstrates the value of communicating and sharing knowledge about what works, what does not work, and why. Successful ICM programs also teach practitioners to integrate, analyze, and adapt knowledge and experience to the needs and contexts of a specific place. CRC developed an action strategy for a key coastal area, Fiji's Coral Coast, to demonstrate how ICM could be implemented to address Fiji's pressing national coastal management issues. CRC used personal coaching and mentoring to build the capacity of individuals and institutions. A national group was then established and mentored to advise and learn from Fiji's Coral Coast demonstration site. That provided a focal point for inter-sector coastal issues and developed a constituency at the national and provincial levels for the development and adoption of a national policy framework for ICM.

In its Community Development Program, CRC worked with local NGOs in the Balik-papan Bay area of East Kalimantan, Indonesia, to strengthen their capacity to implement community-based coastal livelihood and management initiatives. Another small group of ICM-oriented NGOs received assistance in building their organizations (for example, to help them to develop boards and organizational policies, procedures, protocols, and systems) to create a strong institutional underpinning for their primary mission of promoting sustainable resource management.

this system of community-based fisheries management has yielded substantial improvements in harvest efficiency compared with open-access systems (Defeo and Castilla, 2005; Box. 3.2).

Capacity-building will be more successful when it builds on existing capacity, as in the example of the community-based marine protected area of Isla Natividad in México, where local divers were trained by an NGO to carry out more technical ecological monitoring dives and where the local fishers cooperative was already committed to the use of closures to protect shellfish habitats (Box 3.3).

Cooperatives are among the local-level institutions that provide capacity for ocean and coastal stewardship and that can benefit from efforts to "grow" institutional capacities. Existing cooperatives are often organized around marketing and technology, but they offer governance experience, infrastructure, social capital, and environmental knowledge that can be used to enhance ocean and coastal stewardship. The fishing cooperatives of the northwest coast of Baja California Sur (Box 3.3) may be useful models for improving governance. The cooperatives have been able to obtain exclusive fishing concessions from the government and certification for sustainable fishing practices from the international MSC, working with local and international NGOs and government agencies (see Chapter 6).

Local levels of governance are not necessarily on a small scale when urban areas are involved, and urbanization of coastal regions is a major trend. Local governments and cities clearly affect the world's oceans and coasts. Efforts to reform urban develop-

---

Box 3.2
Capacity-Building for the Comanagement of Chilean Coastal Fisheries

The implementation of a national policy to achieve sustainable exploitation levels by restricting access to areas of the coastal seabed in Chile required a substantial investment in capacity-building for the training of fishers, technicians (such as divers and marine technical personnel), and graduate and undergraduate students who contributed to the scientific knowledge base.

The Chilean Fisheries and Aquaculture Law of 1991 defines artisanal fishers and incorporates new regulations that affect user rights through three management components: allocation of exclusive fishing rights within five nautical miles of the shoreline to artisanal fishers, restriction of artisanal fishers' access to the coastal zone adjacent to their regions of residence, and allocation of exclusive benthic-resource extractive rights in given areas of the seabed to organized unions, associations, cooperatives, and registered artisanal fishers.

It took about 20 years to complete the full development and implementation of small-scale comanagement policies regarding the benthic resources of a subset of some 250 small-scale fishers in central Chile. During that time, the total investment was US$5.5 million, which supported the training of roughly 300 small-scale fishers, 28 technicians (such as divers and marine technical personnel), 4 doctoral and 2 masters candidates, and 34 undergraduate students. Researchers in central Chile studied the fisheries and ecosystems to understand restocking rates of benthic resources after area closures and facilitated the legal institutionalization of exclusive territorial user rights for fisheries for two artisanal associations (Caleta Quintay and Caleta el Quisco) and a no-take reserve (Estación Costera de Investigaciones Marinas in Las Cruces), which was established in 1982. This small initial project was the basis of expansion to more than 500 MEABRs, including more than 15,000 fishers along the Chilean coast. The measures have increased fishing income, retained and enhanced community and cultural identity, and served as a basis of community empowerment.

Four elements contributed to the success of the effort:

- The existence of a well-organized system of artisanal fishing communities and national artisanal fisheries associations.
- Successes in the experimental pilot cases and the ability of fishing communities to replicate the successful examples.
- A clear set of rules and the existence of local know-how and technical capacities to expand the implementation of MEABRs.
- Research and publications by scientists who received substantial financial resources from Chile and abroad.

Box 3.3
Marine Reserve Pilot Project in Isla Natividad,
Baja California Sur, México

In August 2006, a marine reserve pilot project was begun in the waters of Isla Natividad through a partnership between a Mexican environmental group, Comunidad y Biodiversidad (COBI), and a fisheries cooperative, the Cooperative Society of Fishing Production Divers and Fishermen of Isla Natividad (Sociedad Cooperativa de Producción Pesquera Buzos y Pescadores de Isla Natividad). The cooperative had made a commitment to sustainable fisheries management, joining a federation of cooperatives that received Marine Stewardship Council (MSC) certification for their lobster fisheries in April 2004. This cooperative had even experimented with closed areas to protect shellfish, hired its own biologists, and set up routine monitoring of lobster and abalone stocks for management purposes.

Isla Natividad, a fully protected marine reserve (that is, closed to all fishing and tourism), is a pilot program handled as field experiments with controls and scientifically designed monitoring to provide data that can be used to decide whether benefits of closures warrant continuation at the end of a 6-year agreement. The cooperative will use the results to help to decide whether to include fully protected marine reserves as part of its management strategy in the future (Comunidad y Biodiversidad, 2006).

Scientific partnerships are an important aspect of the effort. COBI engaged the Partnership for Interdisciplinary Studies of Coastal Oceans (PISCO), a four-university consortium with substantial expertise in marine ecology and oceanography working on the California Current ecosystems of the U.S. west coast. The project is highly participatory, and local fishers perform much of the scientific monitoring. Equally important for the future success of the project is the fact that the program was developed by a local institution. The assessment and evaluation part of the program will include anthropologic exploration of incentives, motivations, social relations, and legal and cultural frameworks that influence this level of engagement by the cooperative and its members and by the partnerships (Comunidad y Biodiversidad, 2006).

ment patterns in the last 20 years have focused on the concept of "sustainable cities" (Cities Alliance, 2007; Smart Growth Network, 2007; United Nations Human Settlement Programme, 2007), integrating sound ecological practices with continuing economic development. The move toward sustainability of urban environments has been coupled with the decentralization of urban governance. Such decentralization is challenging cities' traditional tools, knowledge, skills, and attitudes to implement new governance structures and the management of city services for sustainability. Consequently, capacity-building efforts are also being directed toward political officials, midlevel managers, and others in the cities (Harris, 2006).

Regional Capacity-Building

Large investments in institutional capacity-building are also being made on regional scales, although this geographic scaling up is a somewhat new development. The Partnerships in Environmental Management for the Seas of East Asia (PEMSEA) provides an example of a transboundary program that focuses on partnerships among countries in a broadly defined region to ensure resource sustainability (Box 3.4). A recent report sponsored by the World Bank calls for institutional capacity-building at the regional level and examines potentials for and constraints on regional scaling-up of marine management through marine protected areas (The World Bank, 2006). One of the relevant findings in the report is that marine protected areas are critical in addressing particular challenges of ocean and coastal resource management; however, other instruments may offer more cost-effective and socially acceptable options for scaling up effective marine management. Other instruments cited include MEABRs in Chile (Defeo and Castilla, 2005) and collaborative management areas in Tanzania (The World Bank, 2006).

Recently, regional investments have been made to address issues within the frameworks of large marine ecosystems (LMEs). LMEs are regions of the ocean and coast that include watersheds, river basins, and estuaries and extend seaward to the boundary of continental shelves and margins of coastal current systems. LMEs are delineated according to continuities in their physical and biological characteristics, including bathymetry, substrate composition, hydrography, productivity, and trophically dependent populations. An LME creates an organizational unit for management and governance strategies and so improves coordination of approaches on larger-scale biological and physical processes to reduce fragmentation of data collection and analysis and improve regional decision-making.

The LME approach was developed by IUCN, IOC and other United Nations agencies, and the U.S. National Oceanic and Atmospheric Administration to build capacity and implement programs in the following five categories:

- Productivity—To measure the spatial and temporal distribution of temperature, salinity, oxygen, nutrients, primary productivity, chlorophyll, zooplankton biomass, and aspects of biodiversity.
- Fish and fisheries—To monitor catch and effort, conduct demersal and pelagic fish surveys, measure demographics of fish species, and conduct stock assessments.
- Pollution and ecosystem health—To measure indicators of quality of water, sediments, benthos, and habitats and indicators of fish tissue contamination.

Box 3.4
Partnerships in Environmental Management for the Seas of East Asia

Transboundary management issues have been acute in the seas of East Asia. The area bounded by Brunei Darussalam, Cambodia, the People's Republic of China, the Democratic People's Republic of Korea (North Korea), Indonesia, Japan, Malaysia, the Philippines, the Republic of Korea (South Korea), Singapore, Thailand, and Vietnam has been under intense environmental pressure. The degradation of the region has affected the social structures and economies of the region while depleting resources and affecting human health. Because multiple jurisdictions contribute to the scale of environmental degradation, no single government could be successful in fixing the problems.

PEMSEA was established through the United Nations Development Programme/ Global Environment Facility as the mechanism of intergovernmental cooperation to sustain the natural, sociologic, and economic vitality of the region and to reverse trends of environmental degradation.

PEMSEA emphasizes a holistic, integrative approach to regional governance of the environment through integrated coastal zone management processes and risk-assessment procedures. It initiates networking between local governments to facilitate expanded capacity-building in the region. Through those efforts, the countries of the region can develop and expand their intellectual capital and educate the public about their role as stewards of their environment.

This successful program required the political will of the constituent countries and an influx of political, monetary, and human capital into the program. Multiple political and social components, including NGOs, are partners in PEMSEA. The 2006 evaluation of PEMSEA (Partnerships in Environmental Management for the Seas of East Asia, 2006) concludes that it is "a success worthy of close analysis and possible replication." The committee recognizes PEMSEA as a model for the development and implementation of a cooperative program to advance capacity-building, ensure regional security, and manage and sustain local resources.

- Socioeconomics—To measure economic benefits and costs and social effects of uses of LMEs (Sutinen, 2000).
- Governance—To foster governance institutions to manage uses of LMEs (Juda and Hennessey, 2001).[4]

Regional Seas programs, under the auspices of the United Nations Environment Programme (UNEP), are another large-scale example of capacity-building for ocean and coastal management. Regional institutional capacity-building for the Western Indian Ocean is an example within the framework of the 1985 UNEP Regional Seas Convention. A major endeavor was the creation of Marine Science for Management, a science associa-

---

[4]A handbook was published on governance and socioeconomics of large marine ecosystems by Olsen et al. (2006a).

tion supported mainly by the Swedish Agency for International Development Cooperation (2007) and designed to enhance local scientific capacity for ocean and coastal management. Although it is generally viewed as a success, the problems encountered by the program underscore the difficulty and importance of gaining full participation from each partner country; the effects of differences in political, legal, and institutional frameworks; and the challenges of developing truly sustainable local capacity.

Capacity-Building on a Global Scale

Many international organizations and governments are working toward the establishment of a global system for earth observations that will be networked, integrated, and used for societal benefits. The Group on Earth Observations has the mandate to implement a global system of earth observations; ocean observations would be an integral component. International capacity-building is viewed as a necessary step toward implementation and effective use of global-scale ocean observation systems; many international groups—such as the Partnership for Observation of the Global Oceans (POGO), the Scientific Committee on Oceanic Research (SCOR), and the IOC—are establishing programs to enhance participation of scientists in developing nations. Several specific capacity-building programs are described below.

SCOR contributes to capacity-building by ensuring that every SCOR working group includes scientists in developing countries and countries whose economies are in transition. SCOR provides travel support for scientists in economically disadvantaged nations to attend international scientific meetings through a grant from the U.S. National Science Foundation. Some of the funding is used to support longer-term courses or fellowships, including a program of visiting fellowships for oceanographic observations through POGO. SCOR is developing a capacity-building activity to foster the establishment of regional centers of excellence in marine-science education in Southeast Asia, South America, Africa, and South Asia. The regional graduate schools of oceanography and marine environmental sciences would organize a variety of regional activities in graduate education, bringing together national resources to meet regional needs (Scientific Committee on Oceanic Research, 2007).

The Nippon Foundation and the POGO Visiting Professorship Programme send eminent oceanographers to institutions in developing countries so that they can provide on-site training and mentoring. The goal is for visiting professors and trainees to study and conduct field work together and so foster the exchange of ideas and knowledge. Sustained contacts between trainer and trainees are encouraged even after the formal program has come to an end, and there are plans for evaluating the effects on the trainees' careers.

In 2007, POGO, in cooperation with the Nippon Foundation, announced a new project to establish a Centre of Excellence in Oceanography at one of the POGO member institutes to provide training for oceanographers in developing countries (Partnership for Observation of the Global Oceans, 2007).

POGO promotes opportunities for joint activities between developed and developing countries' institutions as another avenue toward capacity-building. For example, in response to the POGO call for increased observations in the undersampled waters of the Southern Hemisphere, the Japan Agency for Marine-Earth Science and Technology (JAMSTEC) organized a southern circumpolar cruise. JAMSTEC made it a policy to provide opportunities for participation of scientists in developing countries of the region. In addition, JAMSTEC reserved some berths on each leg of the cruise for trainees from developing countries. The onboard training program was implemented by POGO and the International Ocean Colour Coordinating Group with additional support from the IOC.

The IOC's capacity-building effort has extensive programs to train the leaders of oceanographic institutions in developing countries through a series of workshops and meetings. The goal is to identify regional needs and promote sound leadership. IOC also has a floating university program that provides opportunities for scientists to learn how to make scientific measurements at sea and to analyze and apply oceanographic data.

FINDINGS AND RECOMMENDATIONS

Capacity-building efforts are supported by donors who are motivated to build capacity for responsible stewardship. Successful efforts are conducted by dedicated and capable doers in association with highly motivated practitioners who are intent on improving their communities. However, there is a lack of comprehensive information on investments in and outcomes of efforts in capacity-building.

The investment in growing capacity is considerable, but good estimates are not available, because data on capacity-building investments are widely scattered, terminology (such as *capacity-building*) is not used consistently, and many of the data are inaccessible because donors consider them confidential. In addition to information on the magnitude of capacity-building investments, information on the categories of investments (such as the classification scheme used by the World Bank), the approaches used, participation and partnerships, objectives and performance measures, and outcomes would be valuable in developing strategies to enhance the value of future investments.

The committee recommends that donors and doers standardize data collection and analysis to allow the comparison of outcomes among various programs. Such comparisons would increase understanding of the mechanisms and drivers of capacity-building efforts and would help donors and doers to develop more successful programs. At a high-level summit meeting (see chapters 4 and 7) or in another venue, metrics should be collectively established to categorize and assess the wide array of capacity-building processes (for example, differentiating between human or institutional resources and physical infrastructure developments). Investments should be cataloged according to those capacity-building categories to facilitate analysis of program components and comparison of outcomes among programs.

Assessing outcomes is similarly plagued by the lack of information, particularly about governance. The committee recommends that donors sponsor an effort to use case studies in this regard. Case studies are valuable for planning and outreach—they illustrate the lessons learned from both successes and failures. Sets of case studies should be commissioned for each region, including an analysis of relevant governance structures. The case studies should be designed to invite documentation and comparison. The processes and outcomes of governance need to be examined in relation to changes in the condition of the ecosystem (both its human and its environmental components). Such case studies should examine both success and failure in ocean and coastal governance initiatives and should be used as teaching tools in training and university-based curricula and as a means of encouraging transparency and accountability.

4
MOVING TOWARD EFFECTIVENESS: IDENTIFYING BARRIERS TO AND CONSTRAINTS ON EFFECTIVE CAPACITY-BUILDING

HIGHLIGHTS

This chapter:

- Summarizes the principal barriers to and constraints on growing the capacity that will be required for effective stewardship of our oceans and coasts.
- Identifies key principles for future capacity-building efforts.

BARRIERS TO AND CONSTRAINTS ON CAPACITY-BUILDING

Improving the current system for growing capacity for stewardship of ocean and coastal areas requires a critical look at past activities. This chapter identifies the barriers and constraints that have hampered earlier efforts to build resource management capabilities. They range from the inherent complexity of human and natural systems to specific problems in the design, implementation, and scale of aid and development programs. Fragmentation of efforts, lack of political will, and insufficient consideration and assessment of the scale of the problems in designing programs are major barriers to the establishment of stronger, more capable institutions for stewardship of ocean and coastal regions.

Fragmentation

Fragmentation refers to the lack of coordination among efforts to improve the science, management, and governance of ocean and coastal resources. If an individual component of this complex system is strengthened without sufficient consideration of its place within the larger enterprise and other capacity-building endeavors, the overall impact on stewardship will probably be small. In *Capacity Building in Africa: An OED Evaluation of World Bank Support* (The World Bank, 2005), fragmentation is identified as one of the four issues that need improvement:

> Most capacity support remains fragmented. Most capacity building support is designed and managed operation by operation. This makes it difficult to capture cross-sector issues, and to learn lessons across operations. Many capacity building activities are founded on inadequate needs assessments and lack appropriate sequencing of measures aimed at institutional and organizational change and individual skill building.

Fragmentation occurs in various ways in the governance of ocean and coastal areas, such as in split jurisdictional responsibilities (for example, for offshore, coastal, and upstream waters) and in overlapping or even conflicting mandates among organizations that are responsible for ocean and coastal areas. The consequences of fragmented capacity-building efforts include missed opportunities to share ideas, tangible assets, and learning experiences and only partial coverage of the needs because doers or donors are unable to address or uninterested in addressing all aspects of capacity-building. For example, a capacity-building effort may provide outstanding educational opportunities for scientists in a developing country without supporting a complementary effort to build scientific institutions in that country—an essential element in the success of the scientists once they have completed their training. Similarly, the capacity for resource management might be developed without sufficient attention to the need for a legal framework or the need to educate and engage the public to build support for new management initiatives.

Often efforts to develop capacity are aimed at particular issues, such as fishery management or coastal zone management, and lead to sectoral fragmentation. The success of an effort in one sector may depend on the quality of management in another. For instance, fisheries management requires not only controls to prevent overfishing but conservation of fish habitats, such as sea grass beds or marshes, to maintain the productivity of the stocks; effective coastal zone management is required to prevent loss of these valuable habitats. Various aspects of capacity development will be specific to any given sector, but complementary efforts among sectors will be required to reduce fragmentation and make progress toward ecosystem-based management.

Lack of Political Will

World leaders at Johannesburg in 2002 for the World Summit on Sustainable Development emphasized the need for political will to implement change and self-reliance to institute sustainable development. In the present report, the term *political will* is used to describe the resolve of individuals or organizations to bring about change to solve environmental problems. Government institutions often falter in their efforts to increase and sustain capacity in ocean and coastal resources management when political will is lacking. Capacity is developed to achieve particular goals, reach particular targets, enable policy reform, or ensure more effective monitoring and enforcement. On the basis of the literature and direct experience, highlighted below, the committee concludes that the development of political will among institutional leaders in government, nongovernmental organizations (NGOs), and private industry and the coordination of capacity-building efforts among practitioners, doers, and donors will be critical for future stewardship of our oceans and coasts.

The presence or absence of political will determines the success of capacity-building initiatives (see review by Mizrahi, 2004). For instance, local efforts to establish a marine protected area will be unlikely to succeed unless the national policy-makers have also bought into the process. The absence of political will at the national level to implement agreements made at the local or regional level can result in long delays and a potential for special interests to subvert the process by obtaining concessions from national officials who may be corrupt or ignorant of the issues and terms of the agreement. In Haiti, for example, a National Environmental Action Plan to reduce the losses from natural disasters was approved after extensive citizen participation in 1999 but failed to be implemented, partly because of lack of political will and weak institutional capacity (Inter-American Development Bank, 2004; National Academy of Public Administration, 2006).

Lack of political will can also trump the use of the best available science in developing policy options. Political pressures influence political will and result in the trade of sustainable-resources policies for short-term political gain. After the December 26, 2004, tsunami, policy-makers in many of the affected nations endorsed the issuance of thousands of new fishing boats to affected families even though it was generally known that fishing effort exceeded the level that could be sustained by the resource (United Nations Development Programme, 2005).

How, then, can political will be mobilized? One way is through the rise of an effective leader; this is discussed in more detail in Chapter 5. Political will can be fostered in a number of other ways. Key among them are development of public opinion expressed in public forums; reports in the mass media; lobbying through technical papers and public campaigns; the influence of local organizations, national and international NGOs, and donors; and international conferences and reports.

One option would be to create a leadership group for capacity-building involving the United Nations Division for Ocean Affairs and the Law of the Sea, the United Nations Environment Programme, the United Nations Food and Agriculture Organization, and the Intergovernmental Oceanographic Commission as representatives of intergovernmental organizations; donor institutions, such as the World Bank and the Global Environment Facility (GEF); representative NGOs involved in capacity-building, such as the World Conservation Union (IUCN) and the World Wildlife Fund; foundations, such as the Gordon and Betty Moore Foundation, the David and Lucile Packard Foundation, and the Nippon Foundation; and international science and coordination organizations, such as the International Academy of Sciences initiated by the U.S. National Academy of Sciences, the International Council for Science, and the Scientific Committee on Oceanic Research.

A summit, similar to the Earth Summit in 1992 and the World Summit on Sustainable Development in 2002, focused specifically on capacity-building and bringing together institutional leaders from all the major sectors representing doers and donors (for example, governments, NGOs and intergovernmental organizations, academia, and the private sector) would create a forum for developing consensus goals and approaches among the participating nations and institutions. The results of the summit could be used to build political will for increasing capacity in coastal nations around the world. Appendix C of this report provides a brief history of major developments in ocean and coastal management and illustrates the influence of past conferences, summits, and reports on the evolution of initiatives to increase capacity for ocean and coastal stewardship.

Targeted campaigns to raise awareness and study tours to expose policy-makers to international best practices are also useful tools in building political will. The Marine and Coastal Environmental Management Project (MACEMP) in Tanzania is an example of the use of this type of approach to develop political will (Box 4.1).

Political will often depends on timing. Policy analysts and lobbyists seize periods before elections as windows of opportunity to push through environmental reform measures. Political will, once fostered, needs to be sustained to ensure continued support for the reform process. Ultimately, it requires dedicated people to mobilize and maintain political will, which is required to establish a self-sustaining program for informed policy-making and management of ocean and coastal resources. Capacity-building to increase public appreciation of the value of sustainable management of ocean and coastal resources creates a base of support for maintaining political will.

Box 4.1
Mobilizing Political Will to Grow Capacity for Ocean and Coastal Management: Experience in Tanzania

More than a decade ago, Tanzania initiated programs in ocean and coastal management and received donor support to develop an integrated coastal management (ICM) strategy and begin pilot projects for sustainable management of ocean and coastal resources. Many of the early pilot initiatives were sporadic and short term. No budgetary resources were allocated by the government to implement the ICM strategy, nor were they allocated for ocean-resource management. Little or no capacity existed to support the governance of valuable fisheries resources in nearshore areas and offshore. The fisheries were mostly open-access; establishment of marine protected areas failed to curb illegal fishing, including dynamiting. The fisheries in the exclusive economic zone were exploited under a licensing regime that was uncoordinated, unmonitored, and largely unregulated. Ocean and coastal resource management did not have high priority in top management discussions on poverty reduction and economic growth. The lack of capacity and demand for greater capacity in ocean and coastal resource management were not as high a priority for policy-makers as the needs of the agriculture, infrastructure, education, and health fields.

During the period 2002–2004, the findings of a series of studies on the issues and opportunities in ocean and coastal areas were spread among key policy-makers in Tanzania. The values of migratory tuna and other species and the ad hoc licensing regime for the fisheries were debated by parliamentarians. The ongoing press coverage of the need for better management of marine fisheries resources and the plight of the coastal residents, who were among the poorest in the nation, contributed to the mobilization of political will for growing capacity for governance of ocean and coastal resources. The result was the inclusion of the MACEMP in the government's development agenda. The government sought and received support from the World Bank and the GEF for implementing the project. Key categories of capacity that will be built with support from this program include governance of fisheries in the exclusive economic zone, governance of the marine environment in nearshore areas, and support of and services for coastal communities to improve management of the coastal environment. The program will build capacity at all levels (national, district, and community) for better management of resources to add value to the resources harvested, to develop public and private partnerships, and to market the products better.

Corruption

Corruption and mismanagement are major barriers to effective growth of local and national capacity among practitioners. New policies and reforms will be only as effective as the government responsible for implementation and enforcement. There is little incentive for stakeholders to develop the capacity for better ecosystem and resource management if their efforts are likely to be undermined by a corrupt or weak national government.

Corruption can be difficult to overcome. Doers and donors are often negligent in recognizing and acknowledging its existence. Fragmentation contributes to the problem because in many programs the focus is too narrow to address corruption that occurs at higher levels of government—a situation often encountered for the management of ocean and coastal areas. Doers and donors are only now beginning to identify mechanisms for reducing corruption and encouraging transparency. The World Bank, for instance, started World Bank Sanctions Reform in 2005 to uncover and tackle corruption, using new sanctions as an enforcement measure. The bank's anticorruption guidelines, issued in 2006, serve as a legal tool to enable borrowers and recipients to prevent fraud and corruption in projects funded and supported by the World Bank. Under the new policy, the World Bank can sanction persons and entities involved in bank-financed projects that have engaged in defined forms of fraud, corruption, collusion, coercion, or obstruction. Outside the World Bank's efforts, other donors are strategically investing in civil institutions that work to track and uncover waste and fraud; more of this support clearly is needed (U.S. Agency for International Development, 2006).

Issues of Scale

Successfully addressing coastal environmental problems requires recognition of the problems, mobilization of resources to develop solutions, and leadership to drive change (Agardy, 2005). Conceptually, they are best addressed by "thinking globally, acting locally." However, ocean and coastal issues themselves are rarely local in scale, and piecemeal attempts to address them typically fail. The lack of capacity to address large-scale transboundary problems, beyond the small-scale conservation projects and piecemeal ICM efforts, poses a serious challenge to efforts to reverse the environmental degradation that is occurring in all the world's oceans.

Many environmental issues—such as pollution, climate change, protection of ocean and freshwater resources, and biodiversity conservation—are transboundary issues that require multinational government actions or coordinated actions among smaller states. That is the case particularly in the marine context: when resources are shared by more than one country or when consequences result from geographically removed actions, national action alone will not suffice (Kimball, 2001). Many marine species roam across the maritime boundaries of countries, and this places their regulation beyond the control and responsibility of individual countries. In addition, vast areas of the ocean do not fall under the jurisdiction of any nation. The "high seas" are a global commons that cannot be addressed other than through international cooperation and global treaties (de Fontaubert, 2001).

The problem described above is exacerbated by the tendency to invest in capacity independently for nearshore coastal areas and for offshore ocean areas. In nearshore coastal areas, capacity-building activities have tended to be on a local scale (often centered on nearshore protected areas) or limited to a particular sector (focusing, for instance, on water quality and not fisheries or on scientific research and not management). In contrast, capacity-building for understanding and participating in international law, scientific research, and management of offshore marine areas has occurred on regional and even global scales and has been focused largely on larger oceanographic processes and fisheries. There is still a disconnect between the two realms; few capacity-building initiatives seek to bridge the gap between ocean and coastal management.

Challenges Posed by Dynamic Natural and Social Systems

Compounding the inherent complexity of marine ecosystems, the associated political, social, and economic systems contribute layers of complication. Social systems are constantly in flux—perhaps even more than natural systems. Abrupt nonlinear changes can occur in political systems (for example, in elections and revolutions), social systems (for example, in social preferences), or economic systems (for example, in what is produced or how). Initial changes in political, social, or economic spheres often lead to additional changes through the interaction of components of the socioeconomic and natural systems, generating a complex dynamic. For example, an advance in fishing technology from small open boats to trawlers with acoustic fishfinders could cause a dramatic increase in the number of fish caught. The more rapid exploitation rate could result in catch levels that are unsustainable, causing fish stocks to decline and triggering cascading effects on the marine ecosystem.

A fundamental characteristic of natural and socioeconomic systems is the lag between a perturbation of the system and its effects. Culture and tradition may delay societal responses even in the face of changed environmental circumstances. Fixed investments in equipment and infrastructure make fundamental changes in production or consumption expensive. Adaptation to new conditions may take place as equipment and infrastructure wear out and are replaced with investments better suited to the new environmental state. In many regions, population pressures on limited land and water resources, government policies that impede change, and poor access to information or financial resources make adaptation difficult or slow.

The mismatch between the dynamics of natural systems and human responses to changes compromises society's ability to anticipate and develop adaptation strategies to cope with change. Ecological surprises, such as those brought about by species introductions or removals, illustrate how initially small changes in species richness (often just the addition of one species) can trigger dramatic ecosystem effects with potentially large losses in ecosystem services. Therefore, incomplete ecological understanding and the corollary of incomplete sociological understanding can be a major constraint on effective management (Millennium Ecosystem Assessment, 2005b, c, d).

As mentioned in Chapter 2, effective stewardship of dynamic natural and social systems requires multidisciplinary capacities, from specific science-based disciplinary knowledge to an understanding of resource management issues and the conflicting priorities associated with resource use. Capacity-building needs to focus on growing flexible and adaptable tools, knowledge, skills, and attitudes to manage a world of constant change.

Ineffective Ecosystem Governance Structures

Failures in governance impede the growth of capacity. Especially in impoverished, developing countries, the lack of effective governance is a critical barrier to building capacity. Governance is not analogous to "government" but is a composite of the influences that government, civil society, and markets all exert on individuals and societies (see chapters 1 and 5). No single appropriate mode of governance describes all societies and circumstances. Where government is weak, other institutions, such as markets and units of civil society, could provide the leadership to facilitate resource stewardship.

Those concerned with building or growing local capacity should understand the governance situation in the places where they invest, including not only the roles, responsibilities, and strengths of various institutions but the larger societal context in which the institutions are embedded. To improve governance, experts need to identify key leverage points, including investing in leadership development in government, civil society, and the private sector or business community. The issue of governance is discussed in more detail in chapters 5 and 6.

Absence of Horizontal and Vertical Linkages of Ecosystem Governance Structures

Ecosystems and governance occur on a variety of interconnected scales and levels of a governance hierarchy. Although for some purposes it is appropriate to treat a coastal estuary or embayment as an ecosystem, such water bodies exchange with coastal and offshore ecosystems and require a broader approach for some issues. Similarly, coastal communities can be effective stewards for some local aspects of their ocean and coastal areas, but other problems may depend on governance by nearby communities. For offshore areas, typically the national government is responsible.

Some efforts aim to build local capacity, such as the creation of a plan for integrated ocean and coastal management (Payoyo, 1994; Cicin-Sain and Knecht, 1998) or marine protected areas. However, integrated ocean and coastal management may fail unless it is given legitimacy by a wider legal framework. Thus, capacity needs to be built both to develop a plan at the local level and to establish legal institutions to provide legitimacy and authority. Conversely, national legislation or international agreements will not accomplish their objectives unless there is local capacity for implementation.

Capacity-building efforts too often focus on a single ecosystem scale or level in a governance hierarchy without appropriate linkages both horizontally (for example, community to community) and vertically (local to national to regional). Creating a more effective, integrated management system requires attention to the design and implementation of strategies that link local and national activities, recognize the transboundary nature of ocean and coastal issues, and foster international cooperation. Capacity-building efforts will need mechanisms for cooperation and networking locally to globally.

Constraints on Building the Capacity of Institutions

Institution-building is directed at the design, assembly, functioning, and strengthening of institutions. Institutions are formalized communities with a dedicated function or common interest, such as government agencies, university programs, international scientific organizations, environmental groups, and other NGOs. There are common systemic problems in the structure and operation of institutions that undertake or support coastal management and stewardship. Institutions rarely have clearly articulated capacity-building goals or plans to achieve them. To articulate and then achieve such goals requires periodic honest appraisals of programs' strengths and weaknesses.

Many institutions fail to identify and use mechanisms to retain the capacity they have developed; they have not established or implemented incentive structures that encourage shared learning and self-perpetuating training. Typically, people trained in a particular aspect of coastal management are promoted up the management hierarchy when they receive their credentials, and the paradoxical result is that their training is not put to use. Such people should be encouraged to pass on what they have learned to apprentices or colleagues who will be filling their posts when they move on to higher management positions (see the section on brain drain in Chapter 3). Poor recordkeeping and poor maintenance of institutional memory also contribute to the lack of effective mechanisms to retain capacity.

In many instances, programs have focused on training individual management professionals or practitioners as opposed to creating greater awareness and competence throughout an institution. Poorly trained trainers have also made it difficult to develop core capacity and to achieve the kind of multiplier effect that training should accomplish. Rather than searching for ways to support continuing professional development, training efforts often have focused on a single technical subject or on a narrow scope of issues. This focus constrains the development of a broader capacity to manage the diverse conditions found on oceans and coasts.

Institutions are often myopic, being more interested in protecting "turf" than in developing synergistic relationships with other practitioners and institutions in the same region or country. A lack of collaboration and regional strategic planning can lead to redundancy, gaps in management coverage, and a lack of agreement on priority-setting needed to steer capacity-building funds toward the most urgent regional projects and programs. There have been few attempts to have institutions that are involved in developing national plans come together in regions to coordinate their efforts and raise funds for capacity-building in a more strategic and cooperative way.

Conflicting Priorities

Effective stewardship of our oceans and coasts occurs at the convergence of the interests of resource users, resource stakeholders, communities, local and national governments, and international agencies. However, the overlap of such varied institutions and interests with different goals and priorities often results in controversy and conflict. Because most large ocean and coastal ecosystems are not contained within areas owned and managed by single authorities, holistic management is commonly hindered by conflicts among goals, objectives, and responsibilities of the various management agencies. Ocean or coastal ecosystems may not match local government, political, or administrative jurisdictions. Resource allocation needs to be structured so that local objectives are achieved without compromising national well-being or causing undue adverse outcomes for other segments of society.

In British Columbia, the conflict regarding expansion of the salmon aquaculture industry illustrates the overlapping responsibilities of provincial and federal governments. Aquaculture opponents were concerned about effects on wild salmon fisheries and environmental degradation, and proponents were interested in the economic development and employment opportunities offered by expansion of the aquaculture industry. Canada's federal government has the mandate to conserve and protect wild salmon and its habitat, but the provincial government of British Columbia has the primary responsibility for management and development of the aquaculture industry, a situation that directed aquaculture opponents to petition the federal government while proponents sought support from the provincial government (Office of the Auditor General of Canada, 2000; Noakes et al., 2003). Conflicts may also arise from the relative costs and benefits of short-term use versus long-term conservation. Recognition and balancing of the short-term and long-term values are other components of capacity-building for sustainable resource use. A common but rarely mentioned issue in capacity-building is the requirement imposed by governments or donors on practitioners and doers to plan and execute capacity-building projects according to complex monitoring and evaluation-based models. The difficulties of complying with complex management plans and bureaucratic procedures required by the government or donors and the cost of staffing and equipment needed to carry out management and monitoring requirements may be overwhelming. Understanding how to create technical tools for assessment takes time, and the development of such tools imposes huge costs on institutions, especially those in less-developed countries. Furthermore, the benchmarks or indicators that are chosen are often not those which are most meaningful but those which are most likely to be met, and this precludes accurate assessment of whether a project is attaining success.

PRINCIPLES OF EFFECTIVE CAPACITY-BUILDING

On the basis of its review of past and current capacity-building efforts, the committee has identified key barriers to and constraints on such efforts. The analysis, which includes substantial contributions from the participants at the workshop in Panamá, led to the identification of six principles of effective capacity-building:

- Build on past and present capacity-building initiatives.
- Undertake comprehensive needs assessments to identify gaps in capacities before designing a new initiative or to refocus an existing initiative.
- Adopt a strategic approach that responds to the fragmentation in past capacity-building efforts and the inefficiencies in many capacity-building investment strategies.
- Seek partnerships among donors and host nations in developing funds and technical support for the design and delivery of capacity-building initiatives.
- Seek to achieve an inclusive and enabling approach to capacity-building.
- Take a long-term perspective on time and resources required to sustain capacity-building initiatives.

Many of the barriers described in this chapter are jeopardizing the success of efforts to improve ocean and coastal management in some of the countries most in need of a path toward sustainable use of the resources on which they depend. To encourage communities to develop better systems for governing the use of ocean and coastal resources, the doer and donor communities need to design capacity-building programs that make better use of their financial and human resources through greater coordination of efforts and a better understanding of the elements required for success.

FINDINGS AND RECOMMENDATIONS

With the increasing exploitation of ocean and coastal resources around the world, there is an increasing need to increase the capacity for stewardship of the resources. Two shortcomings of capacity-building efforts in particular require urgent attention: the lack of political will and fragmentation of efforts. Without political will, capacity-building efforts are unlikely to receive the long-term support required to grow and retain capacity. Capacity-building is usually treated as one ingredient of programmatic efforts on specific topics. In each case, the identification of capacity-building as a critical ingredient is valid but tends to result in fragmented efforts that fail to connect across sectors, disciplines, and regions.

Although there is a trend toward a more systematic approach to ocean and coastal management worldwide, this has not been true of related capacity-building efforts. The various donor organizations and the agencies and institutions receiving funds have limited their scope largely to their mandates and thereby reduced their ability to contribute to the system as a whole and potentially jeopardized the overall success of their programs (National Research Council, 2001; Ward et al., 2002).

A more coordinated and systematic approach to building capacity for ocean and coastal management and sustainable use is needed. Communication, collaboration, and long-term sustainability should be built into capacity-building programs. Fragmentation should be mitigated by donors through efforts to increase interagency cooperation and to forge linkages among institutions at each level of government. Donors should build political will through better communication about the need to invest in cooperative, coordinated institutions and activities.

Current donor investments in education, training, and outreach initiatives are insufficient. Successful projects and programs have increased access to information and maintained the flow of information. Those programs not only train new personnel but also "train the trainers" in both content and appropriate pedagogic techniques. Lag times and inertia in science-based public policy can derail programs if not anticipated. Institutions need help to strengthen governance, education, and awareness both within institutions and in society at large.

Donor investments in capacity should be strategic and take the priority needs of the community into consideration. Both natural sciences and social sciences should be supported. Partnerships between scientifically advanced and endowed institutions and those needing greater capacity should be encouraged. Donor investments in building capacity should be based on stated needs, not on visions imposed by those with capacity on those lacking capacity. Those attempting to build capacity should be aware of issues of scale and should anticipate the need to build strategic networks of institutions at the various levels of government. In many cases, that will entail regional approaches.

Before investments are made, donors should identify the factors that result in degradation of coastal areas and resources and in unsustainable exploitation of living resources, including problems in governance. In addition to identification of the types of capacity that are needed, the ability of an institution to use funds effectively to increase capacity should be assessed. Donors *are* doers in the very real sense that they have to justify their actions and often have to spend money to raise money. Capacity-building has been and probably will continue to be fraught with risk and uncertainty and will require continuing objective assessment to reveal corruption and ineffectiveness.

Donors would benefit themselves and the cause of ocean stewardship if they communicated more effectively about why investment in ocean stewardship is so important. The donor community has an enormous ability to engage and empower stakeholders who ultimately generate the political will to institute changes in management and maintain greater capacity. Long-term financing for capacity-building, either by individual donors or by teams of donors working together, is critical for growing capacity and realizing the benefits through improved stewardship of ocean and coastal environments and resources.

5
WHAT ASPECTS OF CAPACITY-BUILDING NEED MORE EMPHASIS?

HIGHLIGHTS

This chapter:

- Discusses underemphasized but critically important aspects of capacity-building, including governance, monitoring and enforcement, and leadership development.
- Describes the need for sustainability in capacity-building efforts.
- Outlines the importance of effective program assessment, transfer of information, and investment in networks.

At the outset of a project to improve ocean and coastal stewardship, capacity-building may not have top priority for funders, project managers, and community leaders. Often, there is an impulse to take immediate action that will have quick, demonstrable results coupled with an initial under-recognition of the value of growing capacity for sustaining the impact of individual projects. However, time and experience increase incentives to invest in capacity-building as people see the need for longer-term projects and grass-roots efforts. Longer-term projects can rarely be sustained if there is little or no effort to improve the capabilities of the responsible people, from local stakeholders to government officials. Establishment of a lasting program requires addressing, with or without outside assistance, the various, sometimes conflicting, priorities of the stakeholders and the challenges that they face.

NEEDS ASSESSMENTS FOR CAPACITY-BUILDING

The foundation for programs to improve ocean and coastal stewardship and establish ecosystem-based management depends on identification of constraints and assessment of gaps in knowledge and capabilities. Existing capacity (for example, institutional, managerial, scientific, and governance) and high-priority needs can best be determined by combining the perspectives of the capacity-builders (such as governments, donors, and doers) and recipients or practitioners to ensure that the efforts are sensitive to the needs of the specific site in question (whether on a local, national, regional, or global scale), the particular scope of the needs, and the target audience. The success of an effort will depend on the availability of knowledge of the perspectives of the many groups involved in ocean and coastal management before design and implementation to ensure that the program engages and addresses the concerns of the affected communities. Setting clear and realistic goals for capacity-building projects that account for the different incentives of traditional and nontraditional stakeholders is critical for establishing benchmarks of success. The participatory approach helps in the development of programs that are culturally appropriate and that will garner the support of stakeholders.

Needs assessments indicate where capacity is inadequate and can be used to leverage the resources required for increasing capacity. Assessments may identify opportunities to raise public awareness; to perform outreach activities, to link communities to enhance local capacities, and to strengthen education through formal curricula, informal methods, or other nontraditional means. Additional opportunities may center on training and technical assistance to develop the individual capacities of environmental assessment specialists, planners, managers, researchers, and enforcement personnel. The training might include classic project design and management capabilities but also might involve financial planning and management, communication, and conflict resolution and negotiation. Some important opportunities for capacity-building will involve identification and grooming of leaders and continuing support for emerging leaders in ocean and coastal management at all levels. The importance of leadership and of fostering future leaders is discussed in more detail later in this chapter.

SUSTAINING CAPACITY AND CAPACITY-BUILDING EFFORTS

Capacity is grown through the cumulative efforts of practitioners, doers, and donors to develop self-sustaining programs of knowledge-based ocean and coastal ecosystem-based stewardship. It takes a long time to yield the greatest societal benefits and to adapt to the continually changing conditions of ocean and coastal ecosystems. As ocean and coastal stewardship takes root, sustaining capacity and capacity-building programs becomes a major issue. The typical duration of a program may be adequate to initiate better systems for stewardship, but longer periods are often required to establish self-sufficiency; this applies to initiatives on scales ranging from community-based projects to large marine ecosystem programs. Responsibility for ocean and coastal stewardship ultimately rests with the governing institutions that have the incentive, if not the means, to maintain and grow capacity.

Not enough attention is being given to the long-term financing required to implement capacity-building programs and to practice adaptive management over the decades required to effect substantial change (Olsen et al., 2006a). Many promising efforts wither and die when external funding from a donor community or development banks ends. That issue is often discussed by people involved in capacity-building, but there continues to be little information on the financing of capacity-building activities. Most guides to building capacity have concentrated on financing for marine protected areas (World Conservation Union, 2000; World Wildlife Fund, 2004). There is an urgent need for guidelines on sustainable financing for the broader issues of ocean and coastal governance.

Donor Collaboration

The very short-term planning horizons and project life cycles that characterize investments in building capacity are part of the problem. Donors are naturally reluctant to provide open-ended support, preferring short-term catalytic roles. Long-term sustainable financing is rarely built into project design, so efforts to increase capacity are piecemeal and cease abruptly when funding runs out. The sustainability of capacity-building efforts could be improved by the explicit inclusion of an exit strategy in plans funded by governments or donors.

Sometimes, a donor's world view and agenda produce a "donor-driven" program, often unintentionally, that is not embraced by the targeted communities. A more effective role for donors is to support compatible agendas that governments and communities have adopted, that is, "client-driven" agendas. By working through programs developed in concert with the targeted communities and governments, donor-funded initiatives can more readily transition into self-sufficient programs.

Cooperation among donors can add to capacity-building initiatives. Without cooperation, donors may support the same types of programs, and this would result in redundancy, wasted effort, and competition for the same skilled professionals. However, lack of coordination may leave some important issues unaddressed or underfunded. Joint efforts can result in greater efficiency and can reduce transaction costs. After the initial needs assessment, donors, doers, and practitioners can use this information to set priorities to ensure that the donor assistance addresses the higher priority needs. A coordinated effort by donors can assist nations in implementing their obligations under international and regional conventions. For example, at the international level, many governments have adopted the Millennium Development Goals and the targets of the World Summit on Sustainable Development (WSSD). Coordination and cooperation can also help to create public/private partnerships and link capacity-building efforts in coastal management to those of other initiatives, such as poverty reduction, community development, and economic development.

The Mesoamerican Barrier Reef System Project in México, Belize, Honduras, and Guatemala is a notable example of regional coordination among neighboring countries and international organizations (such as the Caribbean Environment Program, the World Bank, and the Inter-American Development Bank), unilateral aid organizations (such as the U.S. Agency for International Development), private foundations (such as the Summit Foundation, the Oak Foundation, Avina, and MarViva), and nongovernmental organizations ([NGOs] such as the World Wildlife Fund, The Nature Conservancy, Conservation International, and the World Conservation Society). The 1997 Tulum Declaration adopted by the heads of state of México, Belize, Honduras, and Guatemala and an initial intergovernmental action plan indicated high-level support for managing the Meso-American Reef as a single ecological unit. In 2005, the various groups participated in a reiterative consultation with many stakeholders to develop the Meso-American Reef Conservation Action Plan, which established high-priority needs for guiding funding decisions by donors. Capacity-building activities occurred at all levels: scientists, government officials, and local communities. The Conservation Action Plan was presented and adopted by the environment ministers, and a regional coordinating body of ministers played an oversight role. The efforts increased collaboration and sharing of information among all participants, promoted synergies, and avoided duplication of efforts.

Economic and Financial Considerations

The success of capacity-building efforts at the community, national, and global levels depends on a broad understanding among stakeholders of the need for preinvestment to achieve the long-term economic benefits of ocean and coastal stewardship. Before stakeholders will engage in the sometimes tedious and expensive process of implementing stewardship projects and building the necessary capacities, they must expect to receive some return in their investment in the form of social capital. For example, rationalizing the use of fisheries could be viewed as a means for economic enhancement. Restricting access to limited resources, such as oil and gas or mineral reserves, could be described as a way to benefit communities through revenue-sharing for the purpose of sustaining higher incomes.

Stakeholders need to understand that for longer-term sustainability goals to be met, access to resources may have to be limited. That is often difficult to communicate and may require financing to compensate resource users in the short term. Self-financing, community-pooled lending funds, and government loan funds could be used to assist people who are often desperate to meet short-term needs. Regulations and institutions developed through capacity-building will be sustained only if there is adequate monitoring, enforcement, and funding. It may be difficult for institutions and the public to justify such funding when faced with other serious short-term needs, so program planning should place high priority on educating stakeholders in the economic and financial realities of ocean and coastal stewardship to build long-term public support.

A basic understanding of how markets work is also critical for sustaining conservation initiatives over the long term. For instance, fish markets are international, and market prices are beyond local community control. Recognizing that helps fishers to choose strategies with less risk and to enhance the value and sustainability of their resource base. Global thinking and local strategies will be required for success in the global marketplace, including an appreciation of how market niches work (for example, green and organic certifications and ecotourism), the concession process by which governing agents allocate access to resources, and how these processes work for and against fishers or other resource users.

Governing institutions play a critical role in smoothing the transition from short-term resource exploitation to development of long-term, sustainable strategies. Governments have financial mechanisms available to compensate for short-term losses of revenue, such as taxes, user charges, borrowing (in the form of bonds and loans), and grants. Many other mechanisms could provide sustainable financing. For example, funding might be sustained by accessing a share of lottery revenues; dedicated revenues from wildlife stamps; tourism-related fees; fees for ecolabeling and certification, for nonrenewable-resource extraction, or for bioprospecting; fishing licenses and fishing-access agreement revenues; fines for illegal activities; campaigns to establish trust funds; and income derived from local enterprises, such as handicrafts and aquatic products. Collaboration between governments and donors is critical to transition from the shorter-term funding provided by the donor community to sustainable financing from governments or market-based mechanisms to ensure long-term support.

Additional Factors That Affect Sustainability

The duration of government and donor funding is not the only factor that determines the sustainability of capacity. Capacity is more likely to be sustained under the following circumstances:

- The capacity that is built has high priority for the recipients.
- The capacity is suitable for the local context in which it is placed. For example, in most developing countries, a small coastal research vessel, which is relatively inexpensive to operate and maintain, would be more suitable than a large oceanographic vessel.
- The developing country has institutions that can make use of the capacity, for example, fishery management institutions that can make use of increased capacity to assess the status of fish stocks to adjust fishing activities.
- There is a "critical mass" of diverse capacities so that professionals are not isolated but have access to the expertise required to use their own capacities effectively. For example, developing expertise in fish nutrition for use in a hatchery will not result in a successful hatchery operation unless there is also access to expertise in fish diseases.
- There is a long-term strategy to replace donor support with support from within the developing country or countries, for example, through government appropriations or user fees.

The Network of Aquaculture Centres in Asia-Pacific (NACA; Box 5.1) is a good example of partnership institutions that are building capacity for responsible aquaculture in harmony with the environment and other industry sectors. NACA is noteworthy because its government members are primarily developing countries working together to help their own people. Discussions are under way to partner with countries in Africa to form a network among developing nations on different continents. NACA has been sustained for more than 30 years—an exception to the typically short duration of capacity-building programs.

THE NEED FOR EFFECTIVE PROGRAM ASSESSMENT

Standardized criteria for program assessment do not exist, either in the donor community or in the marine management community at large. Such criteria would be of value not only for monitoring progress but for comparing outcomes among programs to evaluate the effectiveness of various approaches. Some NGOs have recently moved in that direc-

Box 5.1
Network of Aquaculture Centres in Asia-Pacific

The Network of Aquaculture Centres in Asia-Pacific (NACA), headquartered in Bangkok, Thailand, was initiated by the Food and Agriculture Organization (FAO) of the United Nations in 1975. NACA is an intergovernmental organization that promotes rural development through sustainable aquaculture to improve rural income, increase food production and foreign-exchange earnings, and diversify farm production. It is a partnership among many countries (Australia, Bangladesh, Cambodia, China, the Democratic People's Republic of Korea, the Hong Kong Special Administrative Region, India, Indonesia, the Islamic Republic of Iran, Malaysia, Myanmar, Nepal, Pakistan, the Philippines, Sri Lanka, Thailand, and Vietnam) and organizations, including FAO, the United Nations Development Programme, the Asian Development Bank, the World Bank, the World Organisation for Animal Health, the Mekong River Commission, the World Wide Fund for Nature, the MacArthur Foundation, and the Rockefeller Brothers Fund. The core activities of NACA include:

- Providing education and training.
- Supporting collaborative research and development.
- Developing information and communication networks.
- Establishing policy guidelines and providing support for policies and institutional capacities.
- Managing aquatic-animal health and disease.

Source: Network of Aquaculture Centres in Asia-Pacific, 2007.

tion and provide a few models for how this might be done. An example is The Nature Conservancy's 5-S Framework, which directs conservation planners and practitioners to address the following five issues: selection of Site, the nature of the System, the Stressors, a Situation analysis, and possible Solutions (The Nature Conservancy, 2000). That type of initiative is an encouraging development in the donor community for both large and small projects. How is Your MPA [Marine Protected Area] Doing? (Pomeroy et al., 2004) is another tool developed to help doers to evaluate marine protected area (MPA) management effectiveness.

Resource limitations and intense competition for funds can put pressure on doers to overstate success. As discussed in Chapter 4, governments and donors may be too focused on performance indicators and less cognizant of whether a project is increasing stewardship capacity. A small set of indicators, with flexibility to accommodate adaptive management, is needed to document and analyze trends in the capacity of institutions to develop ecosystem-based management practices. An example of a prudent set of indicators is presented in *Ecosystem-Based Management: Markers for Assessing Progress* (United Nations Environment Programme/Global Programme of Action for the Protection of the Marine Environment from Land-Based Activities, 2006).

Assessments are most useful when they are shared with all interested institutions and provide realistic information on the progress of a project. Few incentives exist for institutions to report performance outcomes fairly, honestly, and openly. Donors can influence reporting practices by linking funding to assessments of outcomes rather than of performance. Objective identification of constraints on success could be used to target funding and overcome some barriers to improving ocean and coastal management practices.

PROFESSIONAL STANDARDS

A major step in capacity-building is the training of professionals who can move societies toward the knowledgeable use and conservation of ocean and coastal resources for current and future generations. The influence of professionals will depend on society's level of trust in and respect for them for advice and services in support of stewardship. Society is assured that many professions' services and advice meet acceptable standards by certification and licensing programs that are implemented by the professionals themselves or by governments. Such programs are referred to as quality-assurance programs. They usually include processes for reviewing performance to ensure adherence to standards, such as are found in medicine, engineering, law, and public accounting.

Some quality-assurance processes apply to the professional advice and services that support ocean and coastal stewardship (for example, educational institutions are often accredited, and some professional societies have certification programs), but they are much more limited than those in other professions on which the public depends heavily for professional expertise (Sissenwine, 2007). That is an issue in developed countries, as well as in developing countries.

Quality assurance of the professions that support stewardship of oceans and coasts would do more than enhance trust in and respect for professional services and advice. It would also improve quality and reduce confusion. From the perspective of people within the profession, it would be a way to consolidate emerging communities of professionals and a way to advance the codification of good practices. It would also serve as a strategy to promote networking and to encourage members of the profession to come together to define what they believe is most relevant and important in what they do and how they do it. It could lead to the emergence of an epistemic community.

INFORMATION FOR DECISION-MAKING

Stewardship of ocean and coastal ecosystems requires numerous complex decisions in the face of a high degree of uncertainty. The decisions are complex because they involve multiple uses, various sources of information, and diverse societal values. The implications of alternative decisions are highly uncertain because of inherent variability in ecosystems, incomplete scientific understanding, and imperfect implementation of decisions. Some of the uncertainty is considered irreducible in that the future state of ecosystems will never be perfectly predictable even with complete scientific understanding. Thus, a high priority should be given to developing capacity to make complex decisions in the face of uncertainty.

Information Transfer

Low levels of education, inadequate awareness of ocean and coastal issues, uneven access to information, and low environmental literacy are clear constraints on improving management in many places. Greater education and awareness of ocean issues, and specifically how oceans and coasts contribute to human well-being, would provide a foundation for building capacity for ocean stewardship. Donors can invest in specific education and outreach initiatives or can underwrite projects and programs that focus on creating or maintaining access to information. Some programs, such as One Laptop per Child (One Laptop per Child Foundation, 2007), show promise of providing inexpensive technology to impoverished regions. The increased availability of inexpensive or free software, such as Open Source (Open Source Initiative, 2007), can further expand the access of developing countries to information, educational tools, and networking. Information technology programs may include support for Web-based information portals, such as that of the World Ocean Observatory (2007), and for projects to provide hardware, install telecommunication lines, or teach people how to access the information on the Internet.

Facilitating the transfer of information is important, but thought must also be given to passing current knowledge on to future generations. In other words, experts need to know how to educate youth, increase general awareness and knowledge, and ensure the transfer of information between generations (see Box 2.1).

Decision-Support Tools

Decision-support tools are evolving rapidly as society comes to grips with the reality of having to make complex decisions based on limited information in an uncertain world. Some of the capacities needed to use and design decision-support tools can be obtained through traditional academic training. Short courses exist or could be developed to provide some decision-support training. However, the field is evolving so rapidly that much of the necessary capacity must be obtained firsthand by working with the people at the forefront of the field.

Decision-support tools include the following:

- Group processes, which are used to engage stakeholders and decision-makers in identifying and quantifying goals, objective, and acceptable levels of risk. For example, analytical hierarchy processes (AHP) is a mathematical decision-making technique that allows consideration of both qualitative and quantitative aspects of complex decisions by reducing decisions into series of one-to-one comparisons and then synthesizes the results. AHP uses the human ability to compare single properties of alternatives. It not only helps decision-makers choose the best alternative but also provides a clear rationale for the choice. Powerful software is readily available to support AHP.

- *Policy-oriented databases and data-synthesis tools.* Such databases contain information on characteristics of the state of ecosystems, including resources, resource-use activity (for example, where fishing occurs and its intensity), and community profiles. Such data apply to the specific situation (in terms of the time, place, and issue) under consideration. Data-synthesis tools, such as geographic information systems, may be used to visualize the information. There are also policy-oriented databases that contain information from other situations that can be used to judge decision with respect to how they might have worked out somewhere else. Such data are sometimes analyzed to estimate variables that cannot be estimates from data on any specific situation (this is referred to as a meta-analysis). Learning from others' experience is a valuable decision support tool.

- *Risk assessments and decision theory.* This is a well-developed field of mathematical statistics that is used to estimate the probability of various outcomes associated with alternative decisions and to identify the optimal decision according to decision-makers' judgments about the severity of adverse outcomes (minimization of a loss function). Risk assessments are produced routinely as input into fishery management decisions so that decision-makers are informed about the probability of reaching management goals. Such assessments should be a key element of future decisions on rebuilding plans for fisheries to fulfill the WSSD commitment to rebuild by 2015.

- *Modeling.* Models are used to consistently and concisely express beliefs about the state and dynamics of systems and to test the beliefs against available information on the system. Multiple models may be plausible depending on the available data.

Models may differ in the form of equations used to describe the dynamics of the system and parameters of the equations. An example of a particularly complete operating model is Atlantis (Smith et al., 2006), developed by Beth Fulton and Anthony Smith (CSIRO, Australia). It characterizes an entire ecosystem, including key elements of the management process, such as implementation uncertainty.

INVESTING IN REGIONAL CENTERS

Complex and dynamic systems, such as occur in marine and coastal systems, need flexible and diverse tools, knowledge, skills, and attitudes for their study and management. Regional centers, both virtual and real, have been established to enhance collaboration, communication, education, shared visions, and intellectual capacities of people and organizations. Sometimes called centers of excellence, they are created to address specific topics, such as coastal zone management.

An effective way to invest in building capacity is to set up regional centers at universities, such as the U.S. National Sea Grant College Program and the university cooperative in the Philippines (see Chapter 3; Wilburn et al., 2007). Regional centers can support longer-term efforts and provide the resources to develop a cadre of well-trained professionals with a thorough knowledge of the culture, traditions, needs, and capabilities of the target groups whose behavior they hope to influence. Extension programs are most effective when they are linked to a supportive institution with complementary capabilities and continuing programs in research and education. The regional center is one of the most effective models to facilitate the growth of capacities in the management of ocean and coastal resources.

That model to catalyze the regional development of capacities might incorporate several research, teaching, and education centers housed in various universities. The definition of a region in this context may parallel a marine biogeographic province, a political grouping of nations, nations that share cultural elements, or existing regional organizations, such as the United Nations Environment Programme Regional Seas Programme (United Nations Environment Programme, 2005). Each center would specialize in various key elements that are important in marine and coastal resource management, such as small-scale fisheries, fisheries technology, environmental and maritime law, port administration, education and outreach or extension, and coastal tourism. The various centers would offer graduate courses leading to advanced academic degrees for students throughout the region and engage in applied research in their specific fields of interest. Individual centers could also offer short specialized courses in timely topics for professionals throughout the region. Placement of each of the centers would be based on existing strengths of university programs and staff according to predetermined criteria.

A key element in the continued growth and wider extension of capacity-building efforts is the development and support of networks (discussed in the next section). The regional centers could form a network through common coordinated projects, journals and publications, Web-based technologies, the formation of professional organizations throughout the region, exchanges of students and professionals, and information exchanges.

INVESTING IN NETWORKS

To attain goals of effective capacity-building, it is critical to invest in networks. Networks and networking are cost-effective and efficient mechanisms for maintaining and building capacity (see Box 5.2). Networking brings like-minded people together and facilitates the sharing of information and potentially the sharing of resources. That reduces duplication of effort, provides recognition of excellence in universities, increases information exchange, and generates incentives for regional cooperation.

Networks may arise from informal connections developed by people with shared experiences, professional affiliations, or associations of practitioners, but they all create and take advantage of social capital. For example, the Third World Academy of Sciences has promoted networking among developing-world scientists who are studying the use of medicinal plants to create new pharmaceuticals. Social capital was built through two workshops that highlighted successful initiatives. Developing-world scientists keep abreast of their colleagues' activities and developments in their field through a regular newsletter circulated among participating organizations (Third World Academy of Sciences, 2004).

Horizontal and vertical linkages within networks can be built along local, national, and regional scales to strengthen and sustain capacity to manage oceans and coasts and to foster stewardship. Horizontally-linked networks help to connect local capacities to manage ocean and coastal uses and thereby create efficiency. They span wide geographic areas and are important models for linking regional centers to the communities in which they operate. Areas that need greater capacity have fewer resources and a deficit of opportunities for practitioners to interact and draw support from each other. Regional centers and the networks they support can provide an antidote to such problems. A network of like-minded people can share information on what works and what does not work, reduce the sense of isolation by building solidarity and a common purpose with other participants, and thereby promote implementation of the most effective practices.

Vertically-linked networks connect people and institutions up and down the hierarchic system of governance—linking local communities and stakeholders to local governance and connecting local governance to the regional, national, and supranational governance

Box 5.2
Antares Network: Large Gains from Small Beginnings

The Antares Network of South America had modest beginnings but is growing into a substantial global effort in capacity-building in marine and coastal marine ecosystems. It began with a training course in Chile. The University of Concepción organizes Austral Summer Institutes every year designed to provide high-level training to Chilean students and scientists in various aspects of marine sciences. In 2001, one such training program in primary production and remote sensing was organized with additional modest financial support from the International Ocean-Colour Coordinating Group (IOCCG) and the Partnership for Observation of the Global Oceans (POGO), which allowed participants from Latin American and other neighboring countries to make use of the training opportunity in Chile. The trainees in the course were mostly young, dynamic, and keen to make good use of the knowledge gained during the course. But their resources at home were modest, and there was pessimism about what could be gained by solitary efforts in their hometowns. The course instructors recommended that they form a network, join forces, and share resources and ideas. That was achieved under the leadership of Vivian Lutz, from Argentina, who agreed to lead the network. In two workshops in Argentina and in Venezuela (with modest support from IOCCG and POGO), the members deliberated on the next steps. At the Argentina meeting, members prepared a proposal, which was submitted successfully to the Inter-American Institute for Global Change Research, for a small grant to support the establishment of a Web site to serve the network; this resulted in a trilingual Web site (Antares Network, 2007).

Members of the network share data at time-series stations around South America. The network has common elements, but its loose-knit structure allows the members to add elements to local nodes that best suit their needs in spanning such diverse topics as harmful algal blooms, aquaculture and fisheries, and global change. Membership in the network is increasing, and colleagues inform one another of training and funding opportunities and share discoveries and results. The network has fostered participation of members in each other's cruises with free sharing of data. The coordination efforts and successes of the modest Antares Network has drawn international attention, and it has been held up as a model by important international bodies and organizations, such as the Global Ocean Observing System and the Group on Earth Observations. A global network of people with similar interests, called ChloroGIN, has now been formed (ChloroGIN, 2007).

entities above it. Facilitating communication throughout the levels of governance helps to create the conditions necessary for transmitting knowledge about marine ecosystems and concepts for improving ocean and coastal management.

Networks for coastal management are available, but most are fragmented by sector. For example, Capacity-Building for Integrated Water Resources Management (Cap-Net) is an international network made up of a partnership of autonomous international, regional, and national institutions and networks committed to capacity-building in the water sector. Its Web site offers numerous training materials, network management tools, and information on various resources and training courses (Cap-Net, 2004). Cap-Net works well as a model for a coastal management network, but its focus is on fresh water. Some networks, such as the Coral Reef Task Force, are working well on ocean and coastal topics, but in general there are not enough networks in the marine realm to support capacity-building.

CONSIDERING ALL ASPECTS OF GOVERNANCE

As discussed in Chapter 1, the committee defines governance as encompassing the values, policies, laws, and institutions by which a set of issues is addressed. The processes of governance are expressed by three mechanisms: governments, markets, and the institutions and arrangements of civil society (Figure 5.1). Those mechanisms interact with one another in complex and dynamic ways, and they can alter patterns of behavior through such measures as the following (Juda, 1999; Juda and Hennessey, 2001):

- Government:
  o Laws and regulations.
  o Taxation and spending policies.
  o Education and outreach.

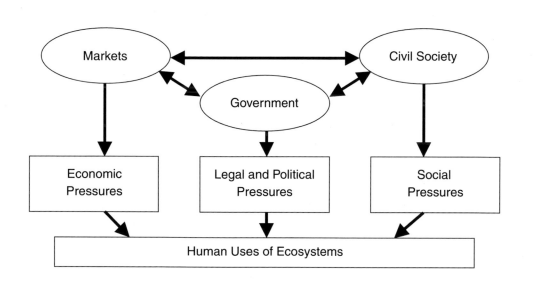

FIGURE 5.1 The three mechanisms of governance (markets, governments, and civil society) and how they are interconnected (modified from Juda and Hennessey, 2001).

- Markets:
  - Profit-seeking.
  - Ecosystem-service evaluation.
  - Ecolabeling and green products.
- Civil society's institutions and arrangements:
  - Vote-casting.
  - Market product choice.
  - Advocacy and lobbying.

COMPONENTS OF EFFECTIVE GOVERNANCE

Meeting the challenges of capacity-building requires the establishment of effective governance structures. The structures must be open, honest, and fair. The global community can provide incentives for good governance through increased trade, monetary rewards, public prizes, and other forms of recognition of leaders, organizations, and communities. Providing such incentives requires an understanding of the nature and structure of governance.

The Pew Oceans Commission (2003) recognized the central importance of the ocean governance system in moving management toward an ecosystem-based approach. All three mechanisms of governance can be used to develop capacity-building efforts that improve long-term stewardship. However, most of the professionals working in ocean and coastal management projects and programs and in international funding agencies have degrees in the natural sciences; few have been exposed to the concepts of governance or to the fundamentals of how markets, governments, and the institutions of civil society function and evolve. That can reduce the effectiveness of otherwise well-educated professionals when they work to apply ecosystem approaches to ocean and coastal governance. The defining features of each governance mechanism and its importance to future capacity-building efforts are described below.

Markets

Markets are institutions that regulate buying and selling activities and may result in the rationing of resources, goods, and services. To function properly, they require preconditions of established property rights and rules of contract and exchange. Markets may or may not operate successfully from society's perspective and may not be functioning when important resources or ecological services need to be rationed. In some instances, it may not be possible to establish and enforce property rights and rules of contract. Markets fail notably when it is not feasible to enforce limited access to a resource, good, or service or where one person's use has beneficial or detrimental effects on others (these are termed externalities), as in the case of watershed deforestation or ocean pollution (Brown, 2001). Markets also fail when a few privileged people have an advantage in the marketplace or when buyers or sellers have inadequate or erroneous information about the contracts or items traded, as in the case of contaminated fish (Hanna, 1998). Failures in the case of ocean governance can be traced back to property-rights issues, enforceability of contracts, uncertainty, and externalities (Edwards, 2005).

Markets and the private sector can be tapped to build capacity for ocean and coastal management and, indeed, stewardship. Although private-sector investment in marine management is overshadowed by government-led management, examples are increasing each day. Box 5.3 highlights private-sector involvement in capacity-building in Chile.

Box 5.3
Private-Industry Involvement in Capacity-Building in the Bioregion of Central Chile

As a condition to receiving a permit to dispose of adequately treated wastewater, a new paper mill, CFI Nueva Aldea, was required to implement a five-year program for development and training in local communities. The resulting program was jointly defined by an assembly of coastal organizations, small-scale fishery villages, local government, the Chilean fisheries authority, CFI Nueva Aldea, associations and unions representing diverse coastal activities, and the University of Concepción.

The program aims to develop livelihood alternatives that meet the following criteria: (1) capacity is built to achieve economic self-sustainability; (2) investment funds are included for infrastructure, training, and education; (3) the success of the program rests not only on the industry but also on the participating individuals and community organizations; and (4) the projects are designed to improve the living standards of the coastal communities. Indicators of performance will be used to identify improvements relative to the baseline documented at the start of the program. Documentation showed limited access to basic services and health care, low levels of education and literacy, a lack of infrastructure for the artisanal fishery, seasonal fishing activities, and high dependence on the coastal ecosystems.

In 2006, the program got under way, and it now includes support for training programs, maintenance of the local community-based coastal fishery comanagement units, renewal and maintenance of fishing gear, commercialization of marine products, continuity of formal education of children and adults, and small-scale businesses, such as aquaculture enterprises and tourism in fishing villages. For most of those activities, state funding has been secured through open competition combined with matching funds from the paper mill.

Although it is too early to evaluate this particular program, private-sector participation in capacity-building could prove to be an effective model because of its potential for attracting additional sponsors. For private industries that make use of land and coastal resources, capacity-building may be an attractive way to catalyze commercial good practices and gain the support of local communities.

Government

Government and civil society play important roles in creating or remedying market failures. Unlike civil society and markets, government can use coercive power to set and enforce rules, as in the case of the recognition and protection of property rights. The legal system, the associated power of adjudication, copyright laws, and the like are all expressions of government. Government can more fully establish and enforce property rights and rules of contract and provide information to buyers and sellers. Government shapes the incentives and disincentives that influence the behavior of the market and provides the rules and services for which the market is not usually an appropriate mechanism. Government can shape the opinions and attitudes of civil society through public education, libraries, Internet sites, and other modes of disseminating information. In democratic systems of government, the public makes government accountable and shapes it actions through voting.

There are many examples of government-led good governance. Sri Lanka is one example of increasing government capacity to promote more effective ecosystem-based management of coastal and marine areas. Sri Lanka dedicated considerable energy to developing a coastal management plan and reorganizing government to carry it to fruition (Lowry and Wickremeratne, 1989; Coast Conservation Department, 1990, 1997; Hale and Kumin, 1992). Another instance of building government capacity is described in Box 5.4.

Civil Society

Civil society consists of the groups and organizations, both formal and informal, that act independently of the government and the market to promote shared interests, purposes, or values in society. The press and associated media may also be important expressions of civil society. Social capital—the informal relations and trust that bring people together to take collective action—forms the fabric of civil society (Coleman, 1988; Putnam, 2000; Pretty, 2003).

Civil society can set the moral context in which markets and government operate. Markets can be harnessed to meet objectives of civil society by, for example, increasing demands for goods and services that reflect good stewardship. Civil society plays crucial roles in ocean and coastal management and strongly affects the ability of both government and commercial interests to practice effective stewardship.

There are many examples of civil-society involvement in ocean and coastal management. Civil societies have participated in all aspects of capacity-building, from education and outreach to training and management. The efforts have occurred on many scales,

Box 5.4
Government Capacity-Building in Darién, Panamá

In 1999, the government of Panamá created and initiated a sustainable development program in Darién Province in eastern Panamá called the Program for the Sustainable Development of Darién (PDSD). Funding for the program (US$88 million) originated in the Panamanian government and the Inter-American Development Bank. Program goals focused on improving the livelihoods of local residents in a manner consistent with the sustainable use of the region's natural resources. Program components included land-use planning, strengthening of institutions, basic services and transportation, and support of actions that promote sustainable production (Suman, 2005).

The institutional and organizational component involved substantial capacity-building activities that aimed to (1) strengthen the institutional capacities to administer the province's natural resources effectively and efficiently; (2) implement measures to mitigate the impacts of new infrastructure; (3) improve the planning, administrative, and financial capabilities of local governments; and (4) increase the participation of community-based organizations in resource management (Inter-American Development Bank, 1998).

For example, PDSD supported actions to strengthen national agencies and organizations that have a presence in Darién (the National Environment Authority, the Panamanian Maritime Authority, the Ministry of Health, the Ministry of Agricultural Development, the Ministry of Public Works, and the Office of Indigenous Affairs). Strengthening the traditionally weak local governments was an additional focus of the PDSD. Activities included improvement of the local tax administration system, of the organization of the community councils, and of municipal planning capabilities. PDSD also worked with 32 community-based organizations (900 people) to encourage them to obtain legal status, to strengthen their organizational and administrative skills, and to access appropriate government agencies better.

One of the land-use planning activities of PDSD focused on coastal management. At the end of 2002, PDSD initiated a coastal management project in Darién Province. The Panamanian institution with formal authority for coastal management was the Marine and Coastal Resources Directorate of the Panamanian Maritime Authority. With university consultants, the integrated coastal management team conducted a diagnosis of the state of Darién's coastal resources and the coastal communities' use of them. Through consultations with community groups, users, and leaders, the project team identified the priority needs of the coastal residents and the threats and vulnerabilities that coastal resources were experiencing. The resulting Darién coastal management plan developed detailed activities to address the high-priority problems in fisheries, resource conservation, coastal-community health, ecotourism, and institutional coordination. The Panamanian government formally adopted the Darién Coastal Management Plan in 2005 (Suman, 2007).

from local on-the-ground conservation projects to global and regional advocacy of policy reform. Box 5.5 shows one example of civil-society involvement.

In general, strategic building of capacity will be achieved best if all aspects of governance are considered fully, including the conventional role of government and the more unconventional (and often overlooked) roles of markets and civil society.

Box 5.5
Wasini Women's Boardwalk

The Wasini Women's Boardwalk is on Wasini Island adjacent to the Kisite Marine National Park and the Mpunguti National Reserve off the Kenyan coast. The island is bordered by a fringing reef on the seaward side and an extensive mangrove stand on the southwestern end. A steady stream of tourists visits the island daily after snorkeling or diving at the adjacent Kisite Marine National Park, providing an accessible market for the boardwalk. Tourists are charged a small fee (US$1.25) for a boardwalk tour that passes by fossil coral structures and goes through a healthy stand of mangroves with opportunities to view sea birds, fiddler crabs, and mollusks.

The boardwalk, owned by the Wasini Women's Group with permission of the Forestry Department, was completed in 2001 with funding from the Kenya Wildlife Service and Netherlands Wetland Program. Various NGOs, including Pact and the World Conservation Union's Eastern Africa Regional Office, and the U.S. Agency for International Development's Kenya program have provided training in business management, governance, and leadership for the group. The project required that the Wasini Women's Group commit to managing the boardwalk as a group, take responsibility for repairs and maintenance to ensure sustainability, pledge a share of revenues to education, and work to minimize cutting of mangroves for fuel.

The beneficial effects of the project include increased goodwill in the local community, which has led to improved surveillance of the MPA (for example, reporting of poaching activities); increased biodiversity protection of the mangrove forest and reefs around Wasini Island; and effective demonstration to the local communities of the nonextractive uses of marine ecosystems.

LEADERSHIP DEVELOPMENT

Leadership is essential for the success of programs to improve ocean and coastal stewardship. There is a need to access and strengthen leadership and train new leaders, both individual and institutional, from the local to the global level. In part, leaders may be trained by example and experience, and the leadership skills of trainers can be models of behavior. The reach and longevity of a capacity-building program can be increased by developing the leadership base at the start. Leadership often achieves its greatest impact and potential to promote change when individual and institutional leaders act as part of a larger network that shares interests, goals, values, and new ways of thinking and doing. Although individuals and institutions may act locally, they should connect regionally and learn and effect change globally as a networked community of practice.

The need for leadership in changing the course of ocean and coastal stewardship is recognized, but leadership has been difficult to define and characterize. The literature offers a wide array of definitions, theories, and styles of leadership and much debate on whether leaders are "born" or "trained." However, it is more important to determine the qualities and attributes that define what people see, feel, or experience when in the presence of leaders. They include the following characteristics:

- Critical and reflective thinking and a willingness to challenge the status quo and invite inquiry into potential new ways of doing and seeing.
- Ability to see the big picture, as well as the parts and their interrelationships—also known as a system orientation.
- Skillful and honest communication, including listening skills and the ability to speak and write with clarity, vision, and purpose.
- Openness to the diversity of world views and perspectives and ability to in make choices, especially when a decision goes against popular thought or opinion.
- Ethical foundation of word and action to navigate the political arena without susceptibility to corruption.

Identifying potential leaders requires sensitivity to cultural conditions and idiosyncrasies. The development of leadership, even in sectors beyond those directly involved in ocean and coastal management, will create the conditions that enable improved ocean and coastal stewardship.

INCREASING CAPACITY FOR ENFORCEMENT AND MONITORING

Compliance requires a broad understanding and acceptance of the rationale for regulations to prevent the degradation of ocean and coastal environments. Understanding may be developed through participatory and representative decision-making. Compliance also requires effective and fair enforcement to maintain support for restrictive regulations, but this aspect of capacity for coastal management is often overlooked, in part because enforcement is typically the responsibility of government agencies that do not receive development aid or donor support. Because of the lack of resources and political will, enforcement of rules and policies adopted by coastal and fisheries management programs tends to have low priority on the political agenda.

Enforcement does not attract donors to the extent that other, more appealing capacity-building initiatives do. Sometimes, the misuse of funds is a particularly acute problem in enforcement because security interests reduce transparency to the point where corruption goes largely unnoticed. When that happens, donors may hesitate to invest in enforcement and related activities.

Monitoring compliance, performing surveillance, and enforcing regulations are activities that are undertaken through a variety of arrangements among government, users or local communities, NGOs, and academia. The most suitable approach depends in large part on the culture and sociopolitical history of the particular place and on feasibility and cost. Boxes 5.6 and 5.7 contain descriptions of two contrasting options for building capacity to undertake enforcement. The first highlights top-down enforcement in the recently established Asinara Marine Park in Italy. Like that of the Galapagos Marine Reserve, which is funded by a stringent system of user fees collected by the national government, the approach of Asinara Marine Park to surveillance and enforcement is very much top-down: there is virtually no involvement of the local community. By comparison, the approach to surveillance and enforcement at the Banco Chinchorro Biosphere Reserve in México is a comanagement arrangement in which the ministries of the Navy, of Transportation and Communications, and of the Environment work cooperatively with the World Wildlife Fund (WWF) and fishing communities to support adequate enforcement.

FINDINGS AND RECOMMENDATIONS

A needs assessment is essential to identify current gaps in capabilities and evaluate conditions that may have hindered earlier efforts. Investments in capacity-building should be based on a thorough assessment of the site, region, or country. Knowledge of the perspectives of the many groups involved in ocean and coastal management and the participation of these groups in assessing needs is essential to ensure that a program engages and addresses the concerns of the affected communities.

High priority needs to be given to the sustainable financing of long-term capacity-building. Typically, investments in building capacity are based on short-term planning and project life cycles. Donors are naturally reluctant to provide open-ended support, and long-term sustainable financing is rarely built into project design. High priority should also be given to the development of guidelines for sustainable financing of ocean and coastal governance and to identification and development of long-term sustainable financing mechanisms for capacity-building.

Investment in regional education and research centers is effective for supporting capacity-building efforts. Performance assessment of established regional centers should be undertaken to assist in the design of future centers. Donors often require results that can be documented after relatively short periods—two to three years. Consequently, they sometimes overinvest in tangible assets at the expense of other types of capacity-building. There is no standardized set of criteria to help donors to conduct systematic assessments.

Assessments of capacity should be conducted regularly to reveal gaps. Standardized criteria should guide the assessments, but small sets of indicators that are flexible enough

Box 5.6
Asinara Marine Park

Italy's Asinara Marine Park is off the northwestern corner of the island of Sardinia. Although it is one of dozens of marine parks in Italy and throughout the Mediterranean Basin, it is in many ways unique. All waters of the Mediterranean have been extensively explored and used for centuries, but the waters around Asinara approximate wilderness. There is almost no coastal development on the island and very limited fishing and other uses of waters offshore. That is because the island has a long history of being closed for various reasons, with strictly protected no-go zones (Figure 5.2) across a wide swath of its coastal waters. In the second half of the 20th century, Asinara housed a maximum-security penitentiary and so was tightly guarded. Before that, the island was the site of a military concentration camp, and even earlier, a place to quarantine people with highly communicable diseases, such as smallpox.

After deciding to close the prison on Asinara, the Italian government recognized its enormous opportunity to preserve a relatively unspoiled and highly diverse set of ecosystems. The Asinara Marine Park was established by presidential decree in 2002; it encompasses 49 mi of coastline around the island and 41 mi^2 of coastal waters. The island's marine environment was the site of a zoning exercise that presented the Italian government with various options to maximize biodiversity, tourism, and fisheries benefits in various zoning configurations (Villa et al., 2002), and a management plan was later adopted. Surveillance and enforcement are undertaken by both the Italian maritime protection authorities (the equivalent of the U.S. Coast Guard), who occasionally visit the island, and dedicated enforcement patrols based in the nearby town of Stintino, who have instituted routine patrols.

There has been little consultation with local people about the design of the park, the purposes of the regulations, and the anticipated benefits of the conservation plan. Similarly, monitoring of compliance with rules, surveillance, and enforcement are all undertaken without participation of local users, such as tourism operators or fishers. But such top-down enforcement is widely accepted, primarily because Asinara has been off limits for so long, and tourism operators have been quick to recognize the marketing potential of even limited access to one of the best protected sites in the entire Mediterranean Sea.

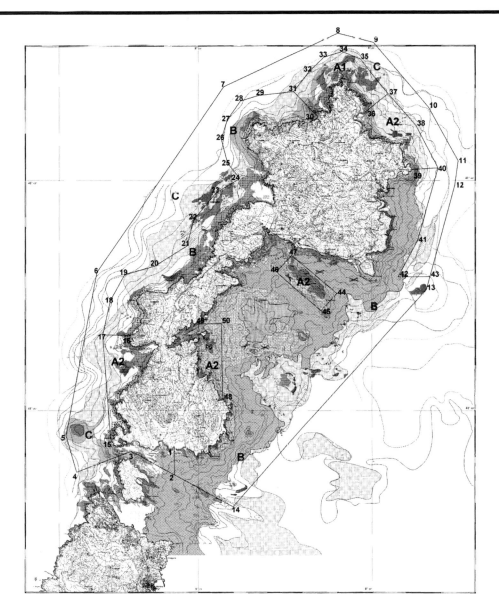

Figure 5.2 No-go zones (denoted by A1, A2, B, and C) in the Asinara Marine Park (reprinted with permission from ICRAM).

Box 5.7
Banco Chinchorro Biosphere Reserve

The Banco Chinchorro Biosphere Reserve is on the southeast coast of the Yucatan Peninsula in México. Its total area is 346,187 acres (140,097 ha) with a geographically unique reef formation consisting of a false atoll with an inner reef lagoon of 143,885 acres (58,228 ha). Diverse ecosystems in the reserve provide shelter and are used as a nursery by several marine and terrestrial species of ecological and commercial importance. Banco Chinchorro is the richest coral reef site in México, with 95 reported coral species, some of which are protected. With its great biodiversity and habitat value, the presence of endemic and threatened species, and relative isolation, Banco Chinchorro is a prime site for conservation and sustainable use.

Such organizations as the National Biodiversity Commission, WWF, and The Nature Conservancy consider Banco Chinchorro an area of high priority. On July 19, 1996, the reserve was declared a natural protected area, and a management plan was developed in 2000. The reserve's main objectives are to ensure the continuity of ecological processes and balance the conservation and use of natural resources through participatory management, scientific research, and environmental education (National Oceanic and Atmospheric Administration, 2007).

The Banco Chinchorro Biosphere Reserve is a prime example of comanagement. The reserve's management plan was developed with the cooperation of fishers and government authorities (such as the Navy Ministry, the Transportation and Communications Ministry, and the Environment Ministry). Licensed fishers support the reserve by providing US$0.20 for each kilogram of conch and lobster they catch. Enforcement and surveillance are supported by WWF, and much of the rest of the reserve is supported by Mexican government funds. The main enforcement objectives are to reduce illegal fishing and to control tourist activities—cruise ships bring 3,000 people per day to the nearby town of Majahual (National Oceanic and Atmospheric Administration, 2007). The result of the comanagement activities has been the elimination of some forms of destructive fishing in the area and the preservation of much of Banco Chinchorro's biodiversity.

to apply to adaptive management also will be required. Reducing the focus on products and outputs and increasing the focus on process and outcomes would enhance the effectiveness of capacity-building efforts. Incentives for fair and honest reporting of progress in capacity-building programs will improve the effectiveness of individual programs and, with more sharing of lessons learned among the donor and doer communities, will advance capacity-building efforts in general.

Donors often require doers to follow complex mechanisms for monitoring and evaluation without giving them the requisite tools. The benchmarks used may not be informative but are selected because they are the most likely to be achieved. That precludes a realistic or accurate assessment of a capacity-building program. The donor community should be active during the evaluation procedure and develop a small set of benchmarks with which doers can accurately assess progress in building stewardship capacity.

Education in and awareness of ocean issues, specifically how oceans and coasts contribute to human well-being, are required to build capacity for ocean stewardship. Donors should underwrite projects and programs that focus on creating access to information and maintaining information networks, including support for Web-based information portals, such as that of the World Ocean Observatory (2007), and support for programs that provide hardware, installation of telecommunication lines, or programs to teach people how to access the information on the Internet. In addition to facilitating the transfer of information, donors should develop strategies to archive current knowledge and transfer it to future generations.

Networks are vital for advancing capacity for ocean and coastal stewardship. Networks and networking are cost-effective and efficient mechanisms for maintaining and building capacity. One of the major benefits of networking is bringing like-minded people together to share information and resources. It avoids duplication of effort, recognizes existing excellence in universities, increases information exchange, and foments regional cooperation.

Networks need to be developed and supported, and their importance needs to be recognized by the doer and donor communities. The focus should be on developing decentralized networks of regional centers that combine research and education with outreach and extension and that foster discussion with nearby populations. It is equally important to foster links between local communities and regional centers and links among local communities, stakeholders, and local governance institutions. A broader network can then be established by linking the regional, national, and supranational governance entities.

Leadership is an underappreciated factor in success. It is critical for the development and sustainability of capacity-building efforts. With their roots in the communities where capacity needs to be increased, leaders can become the cornerstones of successful capacity-building efforts. Leadership comes from a mixture of skills and authenticity, vision and innovation, credibility, compassion and fairness, and determination. Leadership capability is both innate and acquired through training, experience, and mentoring.

Serious investments in developing and supporting leaders should be made at the local, national, regional, and global levels. Donors need to understand, identify, and use leverage points in governance, including investment in leadership development in government, civil society, and the private sector or business community. Training and mentoring leaders, even in sectors beyond those directly involved in ocean and coastal management, will create the talent pool required to enable improvements in ocean and coastal stewardship.

The professional disciplines that support stewardship of oceans and coasts provide services and advice that have direct and substantial effects on society. In addition to developing capacity, there is a need to address professional governance at the level of institutions, education and training, accepted practices, and individual professionals—an issue for both developed countries and developing countries. A suitable model for professional governance should be developed on the basis of some of the lessons learned from other professions. It is time to open a discussion about professional governance that is consistent with the social relevance of the ocean and coastal stewardship professions.

6

BUILDING CAPACITY IN OCEAN AND COASTAL GOVERNANCE

HIGHLIGHTS

This chapter:

- Focuses on aspects of governance related to ecosystem-based management.
- Describes the core tools, knowledge, skills, and attitudes that should be imparted by capacity-building efforts.
- Discusses steps that can be taken to encourage a culture of self-assessment and adaptive management.

Effective and long-lasting ocean and coastal stewardship can occur only when a predictable, efficient, and accountable governance system is in place. There are many descriptions of the phases in which ocean and coastal governance initiatives evolve (Chua and Scura, 1992; United Nations Environment Programme, 1995, 2006; Joint Group of Experts on Scientific Aspects of Marine Environmental Protection, 1996; Cicin-Sain and Knecht, 1998; Olsen et al., 1999; Olsen, 2003; Chua, 2007). The Joint Group of Experts on Scientific Aspects of Marine Environmental Protection (1996) selected the most essential steps, emphasizing that the process is a "cycle of learning" that proceeds from awareness of a set of problems and opportunities to their analysis, formulation of a plan of action, and implementation and evaluation of the plan. Successfully executed governance initiatives establish dynamic processes that are maintained by the active and sustained involvement of the public and stakeholders that have an interest in the allocation of coastal resources and the mediation of conflicts. The various components of governance are described in this chapter with suggestions for infusing knowledge into the practice of capacity-building.

THE RELATIONSHIP BETWEEN SCIENCE AND GOVERNANCE

In the early days of coastal zone management in the United States and of integrated coastal management internationally, it was generally assumed that increasing the scientific underpinnings of planning and decision-making would lead directly to improvements in the stewardship of oceans and coasts. It has since become clear that that was an oversimplified view of the complicated processes of governance that incorporate many, at times conflicting, interests of governments and other stakeholders.

The processes by which options are identified, conflicting interests and values are mediated, and courses of action are negotiated are as important as the application of scientific information (Walters, 1986; Lee, 1993). When conflicts arise during the governance process, opponents often point to scientific uncertainty to justify their positions. Scientific analyses include levels of uncertainty that varies with the complexity and degree of understanding of the system and the amount of information available. Adaptive management, in which uncertainty is reduced through an experimental approach to management, has been recognized as an effective method for managing resources in complex ocean and coastal ecosystems (Hollings, 1978; Walters, 1986; Imperial and Hennessey, 1993; Hennessey, 1994).

GOVERNANCE MECHANISMS AND BUILDING CAPACITY

The power and influence of each of the three governance mechanisms that were discussed in Chapter 5—markets, government, and civil society—fluctuate. But these dynamics are seldom featured in training and advanced-degree programs that prepare people for careers in ocean and coastal conservation, resource management, or development. In most capacity-building programs, it is assumed that government plays the dominant role in ocean and coastal governance. In many regions, however, particularly in developing nations, the power of the government is quite limited, and it is the market—increasingly the global market—that determines how ocean and coastal resources are used and how associated individual development decisions are made. In contexts where informal rules dominate, it is especially important that those working in ocean and coastal management understand the features and dynamics of the three governance mechanisms. Investments in public education and efforts to involve all affected stakeholders in planning and decision-making are most effective when awareness of the likely consequences of different courses of action is complemented by well-informed appreciation of what it takes to use the three governance mechanisms to achieve desired outcomes.

Until the 1990s, ocean and coastal management efforts, particularly in developing countries, focused on developing the roles and responsibilities of the government. Frustration with the many difficulties and recognition of the value of harnessing capability and commitment in civil society led to the channeling of many investments in ocean and coastal management to nongovernmental organizations (NGOs). They also led to programs that involve decentralization of authority and responsibility to local communities and stronger relationships of comanagement between local resource users and government agencies (Olsen and Christie, 2000).

More recently, donors, doers, and practitioners have recognized the power of actively involving business and of working through global markets to improve outcomes. The fair-trade movement, the certification of sustainably fished seafood, the tuna boycott, and the threat of a farmed shrimp boycott show that progress toward improved stewardship can be made by working through market mechanisms. Those well-documented examples of success and failure in the governance of ocean and coastal ecosystems provide valuable lessons for building future programs. Training the capacity-building workforce in the power of market forces will be most persuasive and effective when it draws on case studies and encourages a problem-solving approach rooted in experience in the region of interest.

As discussed in Chapter 4, effective governance requires strong links between the different mechanisms of governance outlined above. If there is no appropriate government policy or regulatory framework for private investment, markets may not work efficiently. Lack of political will to promote a favorable investment climate with a transparent fiscal regime will discourage responsible members of the private sector from investing in sustainable use of ocean and coastal resources. Private investment is necessary to promote growth and reduce poverty in poor coastal nations. Private investors will support regulatory, monitoring, and enforcement initiatives in ocean and coastal areas when it is in their interest. They also invest in capacity-building as an expression of corporate responsibility.

A low government capacity for monitoring and enforcement will constrain the implementation of more sustainable uses of the ecosystem and its resources (see Chapter 5). In the same vein, when civil-society institutions are weak or fail, enforcement of the rule of law becomes a serious challenge for government. Failures of civil society can imperil the security of natural and social resources at local, national, and regional levels and have potential geopolitical effects. The collapse of civil-society organizations and government in Somalia, for example, has increased the environmental stresses on Somalia's ocean and coastal areas, the territorial sea, and the exclusive economic zone of neighboring Kenya. Increasing numbers of migratory fishers and warlords from Somalia are threatening the fishery and other marine resources under Kenyan jurisdiction and creating conflicts in the coastal communities. In addition, markets fail to function well in places where there is an absence of political security because of the failure of government and civil-society organizations.

ASSESSING GOVERNANCE CAPACITY

In Chapter 5, the committee identified the need for developing and using standardized assessment criteria. Here, that discussion is extended from the perspective of how standardized assessments contribute to governance.

Ocean and coastal governance initiatives begin with a series of scoping questions designed to identify major resource management issues, the affected and responsible stakeholders, and the long-term goals of the initiative. A review of the past and current performance of the governance system can be used to establish a governance baseline that complements the initial assessment of resource and ecosystem conditions (Juda and Hennessey, 2001; Olsen et al., 2006b; United Nations Environment Programme, 2006).

It is increasingly unusual for an ocean or coastal management initiative to start with a clean slate. Usually, previous efforts will have either failed to be implemented or had disappointing outcomes. A governance baseline documents the conditions that contributed to previous failures or disappointments and indicates strategies that could improve future efforts. It analyzes the response, or the lack of response, of the governance system to past events and to changes in the ecosystem that are relevant to the initiative. For example, if one issue is overfishing and another is degradation of water quality in an estuary, a governance baseline would show how the three mechanisms of the governance system have responded as contributors or solvers of those two problems. Understanding the traditions and features of the existing governance system requires a long-term perspective on events (see, for example, Putman, 1993).

Establishment of the governance baseline requires an analysis of past and current responses to ecosystem change, identifies the interested parties, and involves people in the affected communities, related businesses, and government. As an initial step, it is useful to develop a chronology of the major events and to examine the effectiveness of the responses. An event might be a collapse of a major commercial fish stock, a decrease in water quality, or a major storm. Experience with this approach has revealed that civil servants in central government departments are often unfamiliar with local events, and community leaders may be unaware of actions and concerns at regional or national levels.

Within an organization, people often differ with their colleagues on whether a particular policy or action is being successfully implemented or whether past decisions have had the desired impact. It is therefore important to bring together stakeholders from different administrations, organizations, agencies, and interest groups to develop the baseline. Additional perspective is gained by soliciting feedback and comments from diverse observers and participants, noting topics on which there is a strong divergence of opinion (United Nations Environment Programme, 2006).

Many years of effort are required to build institutions and governance processes that inspire the trust of and earn commitment from the stakeholders and communities whose support will be needed throughout the stages of the governance cycle. Accountability helps to build trust and requires that the program have clear short-term and long-term goals that address the issues of concern to stakeholders. Useful goals are quantifiable, time-limited, and based on a realistic assessment of existing capacity. Although capacity-building efforts should instill knowledge and skills in the target community, effective governance may also require a shift in attitude toward resource stewardship rather than only resource exploitation; this usually requires many years of sustained effort.

For all those reasons, developing a governance baseline is an important preliminary step to be taken before investments in capacity-building are made. In many cases, baselines reveal shortcomings in governance that could prevent the successful implementation of ecosystem-based management. An ocean and coastal governance baseline indicates where specific, place-by-place adjustments need to be made in the design of future governance initiatives thus avoiding the pitfall of imposing standardized designs and goals on initiatives implemented in different governance contexts. For example, expectations for progress in large marine ecosystem management initiatives, such as those sponsored by the Global Environment Facility (GEF), would be different for the Yellow Sea, the Baltic Sea, and the Benguela Current in large part because the preexisting governance contexts are so dissimilar.

INSTILLING THE TOOLS, KNOWLEDGE, SKILLS, AND ATTITUDES REQUIRED TO PRACTICE ECOSYSTEM-BASED MANAGEMENT

Many tasks are necessary to develop and sustain an ocean and coastal ecosystem management initiative, and they require expertise derived from a variety of disciplines, such as natural science, geology, climatology, economics, law, anthropology, and public education. Typically, professionals are trained in a single discipline and have little exposure to or experience in the other fields. They not only use distinct "languages" but tend to have different world views and values, which are shaped by their educational and professional experience. Stewardship of oceans and coasts requires an ability to integrate diverse perspectives and disciplines. Analysis of the condition and dynamics of an ecosystem, the forces of change, and ecosystem resilience requires a broad knowledge base and the ability to integrate what is known into a framework that addresses problems, builds on opportunities, and takes an area's culture and traditions into consideration. Capacity-building programs will need to instill the tools, knowledge, skills, and attitudes that address the following issues:

- How ecosystems function and change.
- How the processes of governance can influence the trajectories of societal and ecosystem change.
- How strategies can be tailored to the history and culture of a place.
- How to assemble and manage interdisciplinary teams.

Those requirements were discussed in an international symposium, "Educating Coastal Managers" (Crawford et al., 1993), and in later conferences and papers that have reflected on what is being learned from integrating ecosystem-based approaches into ocean and coastal management in developed and developing countries (see, for example, Olsen et al., 1998; Olsen, 2000; Olsen et al., 2006a; Chua, 2007).

The goal of this approach to capacity-building is to generate practitioners of ecosystem-based management who can examine ecosystem processes and responses to change caused by anthropogenic and natural forces. They will need to understand the implications of scientific uncertainty in making management decisions and will need to work with interdisciplinary teams to avoid reactive decision-making and instead formulate effective policies based on the best available science (Pew Oceans Commission, 2003; National Research Council, 2004).

The capabilities required by professionals working toward effective ecosystem stewardship can be taught in a sequence of well-structured training programs or a tertiary degree program using adult learning techniques that rely heavily on case studies. The four major elements of such a future capacity-building program are described below.

Knowledge of How Ocean and Coastal Ecosystems Function and Change

Science provides an understanding of the status and trends of ocean and coastal ecosystems and the causes and consequences of change. Although science can be used as a rationale for particular policy options or lines of argument, it is not the only determinant in the decision-making process. Traditional knowledge and experience-based knowledge contribute to decisions not only as a reflection of local practices and values but also as sources of information when other data are scarce. Practitioners therefore need skills to involve local people to participate in collaborative research and education.

Knowledge of how ocean and coastal ecosystems function and change and particularly of the combinations of conditions that make an ecosystem more or less resilient is necessary to address such critical decisions as the selection of the boundaries for an ecosystem governance initiative. Stresses on marine ecosystems arise from activities at local, national, and international levels that may be driven by market forces, government policies, or cultural traditions. Conflicts between competing activities—such as tourism, industry, fisheries, and recreation—may be amenable to actions on a community scale, but issues posed by freshwater allocation and water pollution typically require consideration of an entire watershed and its associated estuaries. Inasmuch as the great majority of existing policies, plans, and regulations were designed to operate within political boundaries, difficult legal and institutional challenges can be identified and resolved in establishing the boundaries of an ecosystem-based management initiative. It is not a trivial task to select appropriate boundaries for an ocean or coastal management initiative, and the decisions will have a major influence on the eventual effects of the program.

Design and Management of Ecosystem Governance Processes

Designing structures for effective capacity-building requires a holistic perspective that includes understanding the actors who are working within the community and the community structure itself. The skills germane to the design and management of governance systems are discussed below.

Strategic Analysis

The cycle of governance, described in the beginning of this chapter, requires skills in strategic analysis and the policy process. Strategic analysis involves the ability to identify problems and their causes, assess potential solutions, and develop a plan of action. The policy process includes bargaining and negotiation, used to resolve concerns at local, regional, and national levels and to launch new policies. Many of those involved in ocean and coastal management are unfamiliar with the dynamics of the governance cycle and so may be ill prepared to shepherd initiatives through the different stages of planning, winning political support, implementing a plan of action, and evaluating progress. Practitioners of ecosystem-based management require all those skills.

Leadership

Without exception, ocean and coastal management programs that have been shown to be successful have benefited from capable leadership (see Chapter 5). Leaders combine a thorough understanding of how to be an effective actor in governance systems with the technical capabilities necessary to lead a multidisciplinary team. They have the ability to articulate a vision and inspire the collaborative action required to achieve a program's objectives. Leadership, to some degree, is an innate feature that some have and others do not; but leadership is also a skill that can be enhanced through training. The teaching of leadership skills and the mentoring and rewarding of those who demonstrate leadership need to be important features of future capacity-building for ocean and coastal stewardship.

Leaders have the ability to communicate with a multitude of constituents, including donors, scientists, NGOs, and the targeted community itself. The most effective leaders make themselves parts of the communities in which they work. Although it is not necessary that they be from the communities themselves, external leaders may not be accepted by a community, especially if it is known that their presence is only temporary.

Central to the issue of leadership is the recognition that when leaders leave their positions, gaps will naturally develop in the overall structure. The continuing training of new and future leaders is central to the effectiveness of a governance system. Academic and other training programs (see Chapter 3) are critical components of leadership training.

Administration

Administrative skills are essential in the building of complex programs that require collaborative planning and action by diverse government institutions, business interests, scientific organizations, and stakeholders. Negotiating skills are essential because much of the day-to-day business of administering an ocean or coastal management program is dedicated to analyzing and mediating among institutions, groups, and individuals with different interests and diverging values. Such programs need to work to negotiate conflicts and avoid inequitable allocation of coastal resources or degeneration into violent conflict—a situation not uncommon along some coasts (Olsen et al., 1998). As institutions mature, they progress through a predictable sequence of stages, each of which has its own particular features and challenges. The practitioner needs to have an appreciation for this maturation process; this is a common topic in public and business administration.

Public Education and Public Involvement

Public education and involvement of the public in governance processes lie at the center of all successful ocean and coastal management initiatives. In a time of accelerating global change, it is essential to educate the public and stakeholders about the activities that cause changes in ecosystems, the implications of these changes for society, and the options for managing ecosystem effects. Instilling a stewardship ethic in the public is essential for improving ocean and coastal governance, but stewardship cannot be fostered if the public is ignorant or misinformed. Without an acknowledgment of the effects associated with established patterns of behavior and a willingness to take the necessary action, achieving a collective commitment to more responsible lifestyles and new policies will be difficult (Pew Oceans Commission, 2003). Public input is critical for the planning and decision-making process if all interests are to be fairly represented.

Education can make the populace more aware of the finite and fragile condition of the oceans and the destructive aspects of some types of resource extraction and other human activities. The public is less familiar with environmental issues related to the ocean than issues related to land. The linkages between economic sustainability and ecological sustainability need to be more widely appreciated if a stewardship ethic is to take root and flourish (Pew Oceans Commission, 2003). Educational programs for students and adults and for the mass media can help to build a better informed public.

Values and Ethical Dimensions

Many of the effects of human activities are so complex and far-reaching that alternative courses of corrective action may have important ethical implications. They include concerns about equity, with debates over which groups or regions will benefit and which will lose, and the overarching issue of not depriving future generations of the benefits of productive ocean and coastal ecosystems. An effective and responsible leader provides moral leadership, as well as professional competence. The ethical dimensions of ecosystem stewardship need to be addressed explicitly in capacity-building programs designed to generate well-rounded professionals and responsible leaders.

Cultural Literacy

A manager, with much effort and investment, could do well in the skills and knowledge outlined above in the first two categories (how ecosystems function and change and the design and management of governance processes) but still fail as an effective practitioner if he or she does not, or cannot, appreciate the importance of the culture and traditions of the people who are to be served. Cultural literacy—an appreciation of the formal and informal rules of the system, the key players in the system, and the perspectives and constraints of those living within that system—allows programs to be designed to fit the target community. Understanding the local context requires listening to and building on existing expertise and knowledge and recognizing the constraints on growing capacity. Capacity-building programs that do not draw on the experience of the people and places represented by the participants tend to have at best a marginal effect (Olsen, 2000).

A well-prepared professional ideally has a realistic appreciation for the many years of sustained effort necessary to move an ocean or coastal management program from initiation to formal adoption and implementation. Duda (2002) points out that experience in the management of such large ecosystems as the North American Great Lakes, the Baltic Sea, the Rhine basin, and the Mediterranean Sea shows that it took 15–20 years for useful commitments to joint management improvements to be secured from the countries involved. Responses to changes in management take longer in large water bodies stressed by the multiple effects of pollution, overfishing, eutrophication, and habitat alteration caused by the activities of neighboring states. It may take more than 20 years to attain environmental and societal goals in ecosystems on this scale. As a consequence, GEF investments in large marine ecosystem management "will often have ceased before actual water body improvements can be detected" (Duda, 2002). In the case of the Benguela Current Large Marine Ecosystem, the most advanced of the GEF programs supporting large marine ecosystem management, it took seven years to complete the issue analysis and planning process. Transition into an initial phase of implementing the management program formally approved by the participating countries began in 2002.

Access to the Necessary Tools

To be effective, a practitioner of ocean or coastal governance needs access to a large "toolbox" and the knowledge and skills to select and apply the ones that are appropriate for the locale and the issues to be addressed. Personal computers and access to the World Wide Web have in the last two decades revolutionized access to information and to a growing variety of communication mechanisms.[1] Such tools have reduced the isolation that previously created a major barrier for those working in developing countries, in remote regions, and without access to major libraries, sources of data, or professionals with similar interests. Additional tools, such as knowledge management systems and specialized Web sites, are needed to manage the seemingly limitless amount of information that has become available through the Internet.

---

[1]Access require basic infrastructure, such as a reliable source of electricity.

Other tools that are often useful or necessary in an ocean and coastal stewardship initiative are in several broad categories. The first category consists of tools that make it possible to monitor change in key environmental variables. They range from relatively low-technology instruments—such as a thermometer, salinometer, and secchi disk—to sophisticated but increasingly available automatic sampling and recording instruments that can be installed to monitor changes in such fundamental variables as temperature, salinity, nutrient concentrations, and even specific pollutants of concern.

A second category consists of tools that characterize human populations and monitor changes in socioeconomic variables. A wide array of rapid-assessment protocols and surveys that provide information on needs, values, perceptions of issues, and socioeconomic variables are readily available and can be matched to the specific issues, technical capabilities, and funding availability.

The practitioner also needs to be familiar with the toolbox of regulatory and nonregulatory tools that can be applied to influence or regulate human activities. They include land and water zoning schemes, such as a variety of marine protected areas; fishery management tools; permit programs; regulations of many kinds; and the equally important nonregulatory measures, such as incentive programs, investments in capacity-building, and public education techniques.

Geographic information systems (GISs) have become a powerful tool for displaying spatially expressed variables on electronic mapped overlays (see chapters 3 and 5). GISs can be used to visualize and integrate across the environmental, social, economic, and institutional dimensions of an area. GISs are particularly useful in preparing scenarios that illustrate the potential outcomes of different courses of action. An array of decision-support tools can be very useful in integrating the available science and helping to evaluate alternative courses of action.

Yet another category of tools consists of the case studies, simulation exercises, and small-group problem-solving that are valuable in capacity-building and engaging with stakeholders in a given locale. Effective capacity-building activities create familiarity with those tools, including the ability to select and access the most appropriate ones for a given locale at a given stage in the evolution of a stewardship initiative.

Assembling and Managing Interdisciplinary Teams

Effective stewardship requires capacities that are multidisciplinary—incorporating observations of the physical and chemical environment; ecosystem properties, processes, impacts of human activities, and biodiversity. Building capacity entails such factors as human-resource development through education and training, institutional and infrastructure development, and the creation of favorable policy environments that encompass an array of public and private stakeholders. Interdisciplinary teams of experts whose missions include addressing management responses to human uses of the oceans and coasts may focus on the various stages in the sequence of strategic environmental planning. Interdisciplinary teams of professionals draw on the experience and perspectives of a broad spectrum of disciplines to address complex ocean and coastal management issues.

For example, development of a coral reef management plan initially requires characterization of the reef's natural and physical environment (for example, ecology, fish populations, water quality and currents, and spatial relationship to other ecosystems), the human uses and pressures on the reef ecosystem (for example, tourism, fisheries, water pollution from adjacent areas, and effects of global climate change), and the institutional and legal framework (government institutions with authority over various aspects of the reef system—coral colonies, water quality, fisheries, and other human uses). Implementation of a coral reef management plan requires expertise to develop strategies and vehicles to communicate effectively with different groups, enforcement, and public-education campaigns. Natural scientists contribute knowledge of techniques for observing ecosystem changes, and social scientists provide expertise for assessing social impacts and the costs and benefits of the strategies. Thus, solutions to problems will depend on the coordinated participation of specialists in many fields and disciplines.

In many situations, the "nontraditional expert"[2] has important information about the state of a resource, potential environmental and social effects of a development, and effectiveness of actions and strategies. Nontraditional experts may be fishers, community leaders, naturalists, or vessel operators; they may have a stake in the outcome, but their participation in the project team may provide unique and valuable information.

CODIFYING GOOD PRACTICES AND DEVELOPING CERTIFICATION STANDARDS

The practices of ocean and coastal governance have matured to the point where it is possible and useful to codify what has been learned and is emerging as internationally accepted good practices. Good practices may be codified in many ways. The institutions that fund ocean and coastal management programs, for example, have standards and criteria to guide the selection of projects, the monitoring and evaluation of project performance, and the choice of benchmarks of achievement. Similarly, institutions responsible for the implementation of ocean and coastal management programs codify how they will

---

[2]"Nontraditional experts" are people who do not possess academic credentials but have extensive direct experience.

operate and what they require of their partners and those whose activities they oversee or regulate. Those forms of codification are internal to the institutions.

Another approach to codification of good practices is the Code of Conduct for Responsible Fisheries (Food and Agriculture Organization of the United Nations, 2007c). The code has had considerable impact as a benchmark against which to gauge the behavior of individual fisheries worldwide. The Marine Stewardship Council (MSC) is an example of a nongovernment initiative to develop standards for sustainable and well-managed fisheries. MSC was started as a partnership between Unilever, a seafood buyer, and the World Wildlife Fund, an international conservation organization, and is now an independent organization with a broad funding base. MSC took two years to develop its standards based on "worldwide consultation with scientists, fisheries experts, environmental organizations and other people with a strong interest in preserving fish stocks for the future" (Marine Stewardship Council, 2002). The standards are used to encourage responsible fishing practices through market forces by providing consumers with the option to purchase seafood preferentially from MSC-certified fisheries.

One strategy for strengthening capacity for ocean and coastal governance is to develop a professional certification program (see chapters 3 and 5). This approach to the codification of good practice sets explicit standards for professionals and has been found to be beneficial in many professions, including law, medicine, engineering, and the sciences. The professional certification programs of the American Fisheries Society and the Ecological Society of America are well-developed and respected systems for recognizing levels of professional knowledge and competence. The objectives of such professional certification are to provide government and nongovernmental agencies and organizations, private firms, courts, and the general public with standards of experience and education for qualified professionals and to recognize professionals as educated, experienced, and ethical and as acting in the best interest of society and the public. In addition, professional certification promotes and encourages the further development of a field's professional standards. Certification is awarded through peer evaluations of the qualifications of applicants by defining: (1) the qualities of a professional, (2) the educational standards and experiential requirements associated with each level of certification, (3) the ethical standards for the profession, and (4) the standards for continued professional development.

Future capacity-building for ocean and coastal governance could be enhanced by the development of professional certification programs. That could go far toward codifying and promoting wider recognition of the unusual combination of tools, knowledge, skills, and attitudes described in the previous section.

BUILDING A CULTURE OF LEARNING AND SELF-ASSESSMENT: THE BASIS OF ADAPTIVE MANAGEMENT

As mentioned above, ecosystem-based management and ecosystem stewardship often take many years to be established and to yield results. Hence, success depends on the design and implementation of marine environmental programs that can adapt to social and environmental change and can adjust their approach on the basis of experience. All too often, the difficulties of practicing adaptive ocean and coastal management and the short-term and fragmented approaches to both governance and capacity-building (see chapters 3 and 4) have hindered implementation, especially in developing regions. Although ocean and coastal management programs may be good to excellent in technical quality, they typically fall short in implementing their action plans.

The response to that situation in many cases has been to impose increasingly complex monitoring and evaluation systems that consume considerable time and resources and create frustration in all involved. The fundamental problem is that most investments in ocean and coastal governance in developing regions take the form of short-term, often disconnected projects. The results of the infusions of funds, technical assistance, and capacity-building are typically measured by such outputs as reports produced, meetings held, people trained, equipment purchased, and activities completed. However, the connection of such outputs to progress toward the fundamental goals of improved ecosystem and natural resources management is tenuous or absent.

One alternative is results-based management and a demand that projects and programs produce tangible evidence of change in society and in the ecosystems of concern. In the long-term endeavor to establish governance systems for ecosystem-based management, it is necessary to define a sequence of outcomes that signal progress toward goals that can be achieved only over the long term. The "Orders of Outcomes" framework illustrates the approach (Figure 6.1; Olsen, 2003). The ultimate goal of sustainable development is disaggregated into a sequence of tangible levels of achievement. The focus is on outcomes rather than processes. Sets of markers or indicators that can be used to assess progress in an ecosystem-based management initiative are identified. The framework has been applied to programs that seek to integrate management of river basins, coasts, and large marine ecosystems (United Nations Environment Programme, 2006).

The first order is achieved by assembling the enabling conditions for the sustained practice of ecosystem-based management. It culminates in negotiation of commitments to implement a plan of action directed at a set of high-priority management issues. The implementation of a plan of action is addressed in the second order, as changes occur in the behavior of institutions and relevant user groups and as success grows in generating the funds required to sustain a program over the long term. The third order marks the

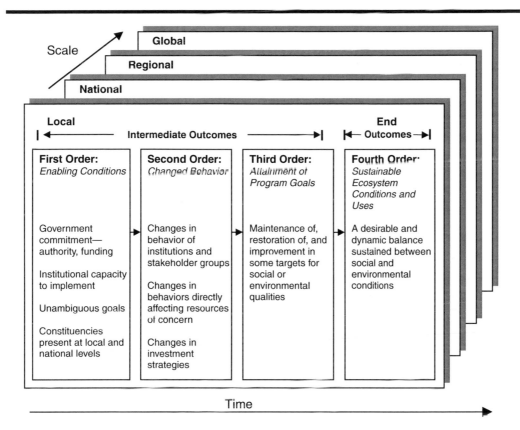

FIGURE 6.1 The four Orders of Outcomes in ecosystem-based management. Source: Modified from Olsen, 2003; reprinted with permission of Elsevier Limited.

achievement of the specific social and environmental quality goals that prompted the entire effort. Such an approach is useful in that it focuses investments in capacity-building on the distinct thresholds of capacity to practice ecosystem-based management relative to a governance baseline. If such a framework were widely applied, it would simplify monitoring and evaluation by focusing attention on the outcomes that are relevant and achievable at a given stage in the evolution of a program and within a defined timeframe.

Periodic self-assessments (see Chapter 5) that draw on the Orders of Outcomes framework and similar integrating heuristics can provide the basis of adaptive management. The objective of self-assessment is to internalize the learning process and encourage adjustments as an initiative matures, responds to its own experience, and adapts to changes in the social, political, and environmental context in which it is operating. Such a culture of self-assessment would constitute a major change in the practice of ocean and coastal governance and mark a substantial increase in the capacity of the organizations involved.

FINDINGS AND RECOMMENDATIONS

A central assumption of past coastal zone management initiatives was that the principle barrier to improving stewardship was the generation of scientific knowledge and its application to planning and decision-making, but barriers also arise in the governance process because changes in management affect the diverse and often conflicting interests and values of people and institutions. Therefore, the governance dimensions of stewardship need to be at the heart of both needs assessments and capacity-building investments.

Capacity-building for ocean and coastal management should seek to improve governance at all levels, targeting civil society and markets in addition to government. No formulaic approach can be universally applied, but a strategic approach that assesses governance baselines and identifies barriers to implementation will provide a sounder basis for designing programs to encourage environmental stewardship.

A governance baseline that describes the strengths and interactions among the three governance mechanisms (markets, government, and civil society) revealed by past and current performance should be established. Capacity-building programs should be tailored to address the specific strengths and weaknesses of a region's ocean and coastal governance system.

Governance baselines and associated capacity-building assessments should be assembled in close consultation with stakeholders drawn from the full array of governance mechanisms and levels, for example, local, state or provincial, and national government; the private sector with economic interests in the ecosystem and its resources; and civil society as represented by academia, NGOs, and social groups. Assessments should be developed, with opportunities for the interested public to offer perspectives, concerns, and recommendations as that is feasible and appropriate. The development of a governance baseline is an important preliminary step to be taken before investments in capacity-building are made.

The committee strongly encourages organizations that are undertaking capacity-building programs to develop common frameworks for analyzing trends and events in ocean and coastal ecosystems and evaluating the responses of the governance system. The Orders of Outcomes framework offers one option for such an integrating system. It can be used to develop parsimonious sets of indicators (see, for example, United Nations Environment Programme, 2006) that gauge the capacity in a given place to practice ecosystem-based management, assess progress in implementing a plan of action as reflected in behavior changes in actors and institutions, and suggest a basis of tracking changes in the condition of an ecosystem that are relevant to the goals of a stewardship initiative. Assessments based on such common frameworks should be readily accessible and widely distributed to encourage the development of a common knowledge base. Shared experiences and successes help to strengthen constituencies for a stewardship ethic and generate the political will required for sustained, purposeful action.

7

THE PATH AHEAD: STRATEGIC AND LONG-TERM APPROACHES TO CAPACITY-BUILDING

A VISION FOR THE FUTURE

The well-known proverb "Give a man a fish, and you feed him for a day; teach a man to fish, and you feed him for a lifetime" often has served as the rallying cry for capacity-building efforts. This proverb emphasizes the importance of teaching skills for self-sufficiency; however, the modern understanding of ocean and coastal ecosystems makes it necessary to revise the proverb and the approach to capacity-building.

Recognition of the current degradation of ocean and coastal ecosystems and the loss of ecosystem services makes it necessary to expand the scope of capacity-building beyond skills for self-sufficiency to include skills for ensuring the long-term sustainability of living resources and other ecosystem services. This will require imparting a more comprehensive understanding of ecosystems and of the direct and indirect effects of human activities, such as fishing, coastal development, waste disposal, and port-dredging. The committee proposes a modern-day proverb for capacity-building: "Give a man a fish, and you feed him for a day; teach a man to fish, and you feed him for a lifetime; teach a man to fish sustainably, and you feed him and his descendents for generations to come."

The committee envisages stewardship based on the recognition of the interconnectedness of human activity and ecosystem health. When investments in capacity-building are designed and delivered as contributions to sustained collaborative efforts to meet the conditions and needs of each community, localized investments in capacity can be complemented by collaborative global programs designed to sustain ocean and coastal governance and management initiatives in a diversity of settings. A global network of ocean and coastal stewardship programs could document progress toward defined goals of ecosystem condition and provide living models that inspire and guide others. Such a network could be a primary source of the materials and long-term case studies used in capacity-building programs worldwide.

How will society know that investments in capacity-building efforts by communities, donors, and doers are advancing those goals? The following actions could be used to assess progress toward the goals and vision discussed in this report:

- **Document changes in capacity through assessments that use a consistent set of criteria.** Regular assessments will be needed to help programs to adapt to changing needs in long-term capacity-building efforts. Some common criteria will facilitate comparisons through time and across programs, but assessments will also need to be tailored to fit the circumstances and characteristics of specific programs.
- **Fund capacity-building through diverse sources and coordinated investments by local, regional, and international donors.** Building sustainable programs requires longer-term support than is typically provided by individual donors.
- **Support dynamic and committed leaders, usually local, to develop a culture of stewardship and to work with the community in the development and implementation of an action plan to sustain or improve ocean and coastal conditions.** Effective leaders also serve as mentors and role models that can motivate future leaders.
- **Develop political will to address ocean and coastal management challenges.** Political will requires building a base of support for ocean and coastal stewardship through greater awareness of its long-term societal benefits. Public discussion of the costs and benefits of environmental sustainability—stimulated by the mass media, information campaigns, and educational programs—will heighten awareness and build political will for necessary changes in the processes of planning and decision-making.
- **Establish continuing-education and certification programs to build the capabilities of practitioners.** This will enable current and future generations of professionals to adapt and apply the best practices for ocean and coastal management in diverse settings.
- **Establish networks of practitioners to increase communication and support ecosystem-based management along coastlines, in estuaries, and in adjoining large**

marine ecosystems and watersheds. The networks will facilitate collection and integration of knowledge, new technologies, and Web-based data management systems in support of locally implemented, regionally effective, ecosystem-based management.

- **Create regional centers, where appropriate, to encourage and support integrated ocean and coastal management through collaboration among programs in neighboring countries.** The centers would link education, research, and extension to address issues of concern in the region, providing an issue-driven, problem-solving approach to capacity-building.

RECOMMENDATIONS

The committee offers several broad recommendations to link investments in capacity-building to on-the-ground improvements in the condition of ocean and coastal ecosystems. The recommendations address the barriers to more effective ocean and coastal stewardship identified during the course of this study.

Base Future Investments in Capacity-Building on Regional Needs Assessments

A recurring theme among experienced practitioners in capacity-building programs to institute ecosystem-based management is the importance of anchoring capacity-building in thorough needs assessments. To establish a baseline, assessments would be most effective if they examined not only environmental, social, and economic conditions but also the existing governance structure (see Chapter 6). Periodic assessments of ocean and coastal governance needs should be conducted on a regional scale. The assessments should cover both indicators of ecosystem change and governance responses to change on a scale that encompasses watersheds, coastal regions, and marine ecosystems.

The pace of ecosystem change and response to change is such that priorities will need to be updated every three to five years on the basis of needs assessments. Periodic assessments will be required for each region because the maturity, capabilities, challenges, and traditions of governance differ from one place to another. Capacity-building priorities in Southeast Asia and the best strategies for meeting them will be quite different from those appropriate for East Africa or Central America.

Each assessment will require consultation with the most relevant stakeholders in government, civil society, and businesses to solicit their views on the strengths and weaknesses of and lessons emerging from past and current initiatives for ocean and coastal governance. The credibility of the periodic needs assessments will depend on the participation and buy-in of the major investors in capacity-building in that region. The latter may include the government agencies most directly involved in ocean and coastal management, the businesses that rely on ocean and coastal resources, bilateral and multilateral donors, representatives of such specialized international institutions as the United Nations Food and Agriculture Organization, and representatives of the major nongovernmental organizations (NGOs) actively involved in ocean and coastal capacity-building in the region. Each assessment should be designed to attract high-level attention and generate interest and visibility in high-priority issues. The findings should form the basis of regional action plans that will guide investments in capacity and that will be implemented with realistic milestones and performance measures. Action plans should include concrete agreements on roles and responsibilities of donors and doers.

Build Capacity to Generate Sustained Funding for Ocean and Coastal Governance

Ecosystem-based management can produce desired outcomes only when it is sustained over the long term. In developing countries, where ocean and coastal change and the loss of critically important goods and services are most rapid, the dominant mode of investment in ecosystem-based management is two to ten year "projects"; most initiatives are funded for five years or less. That applies to initiatives on scales ranging from community-based projects to the large marine ecosystem programs supported by the Global Environment Facility. Many promising efforts wither and die when external funding from the donor community or development banks ends. Little attention is given in capacity-building programs to the long-term financing required to implement programs and practice adaptive management over a period of decades (Olsen et al., 2006a). There is an urgent need to build awareness of this problem so that future programs are designed and implemented with strategies for sustained financing. The guides that have been produced to date (e.g., World Conservation Union, 2000; World Wildlife Fund, 2004) concentrate on financing for protected areas. There is an urgent need for broader financing guidelines to serve the multiple aspects of ocean and coastal governance from local to regional levels. Guidelines should provide practitioners with the knowledge and skills required for applying such market-based mechanisms as user fees, regulatory fees, beneficiary-based taxes, and liability-based taxes.

Develop, Mentor, and Reward Leadership

One of the most important success factors in developing capacity for stewardship of oceans and coasts is leadership. Leadership is important in research, education and training, institutions and governance, and civil society, from the local to the global level. Leadership conveys a shared vision that motivates and empowers people, focuses activities, and provides the confidence to reflect on progress and be adaptive. One strategy to strengthen capacity and stewardship is to identify, develop, mentor, and reward leaders as they emerge in projects and programs or are attracted from the larger society. Leaders are gifted communicators and play a central role in navigating the process of assembling support for a course of action. They may not excel in the technical abilities that are usually the primary criteria for assembling the teams that populate ocean and coastal management initiatives. The capabilities of leaders should be built through programs designed specifically to address their needs. Such investments in leadership will be most effective when they are associated with a regional community of programs and people that are working to achieve common goals.

Support the Creation of Networks

The creation of networks should be encouraged as a way to bring together those working in the same or similar ecosystems with comparable management or governance challenges to share information, pool resources, and learn from one another. Networks are cost-effective and efficient mechanisms for maintaining and building capacity. They foster the creation of epistemic communities based on trust and mutual respect. Well-structured networks help communities to envision the bigger picture and reduce members' sense of isolation by building solidarity and a common purpose with one another. Networks associated with periodic regional assessments of needs and progress can encourage discourse and critical examination regarding what does and does not work and thereby promote implementation of successful practices.

Networks are enhanced by periodic personal contacts, but much can be accomplished through well-structured and adequately maintained Web-based systems. The latter need to be designed to allow users to access information on a given topic in a systematic manner and to incorporate methods for ensuring quality control. Knowledge management systems could provide practitioners with information for analyzing capacity-building successes and failures if they specifically document experiences with ecosystem-based management and the strategies used to overcome the implementation gap. The systems could also document how specific technical and policy issues have been resolved, identify opportunities for transboundary collaboration, and provide access to public education materials and meeting summaries produced by participating programs.

Establish Regional Centers for Ocean and Coastal Stewardship

Regional centers for ocean and coastal stewardship should be established as "primary nodes" for networks that will coalesce efforts to fulfill action plans. The centers would require a contingent of professionals with hands-on experience and infrastructure to serve as a resource for the entire network.

Most effective are decentralized networks and centers that combine research and education with outreach and extension and foster discussion with the surrounding communities. The Land-Grant University System and the National Sea Grant College Program in the United States are only two examples of networks of institutions that have fostered the adoption of new practices in agriculture, aquaculture, public health, and education. Long-term, sustained efforts have been the key to the success of those programs. The adaptation and application of the integrated education-research-extension model on national and regional scales as a primary strategy for developing capacity for ocean and coastal governance would offer a powerful alternative to the current pattern of investment in expensive, short-term, disconnected projects.

Document and Widely Disseminate Progress in Ocean and Coastal Governance

A central feature of periodic needs assessments is that they will document and draw on the evolution over time of selected ecosystem-based stewardship initiatives in each region. It will be particularly important to integrate the often rich but currently scattered information on and experience in ecosystem change and governance initiatives in linked watersheds, estuaries, and large marine ecosystems. Where appropriate, such analysis should showcase the successful application of ecosystem-based management principles and practices through the societal and environmental benefits generated. Such sustained long-term documentation and analysis of ecosystem-based management initiatives in diverse cultural, geographic, and biophysical settings will be of great value to future capacity-building and could also help to build political will for integrating approaches into planning and decision-making. Documentation should cover a common conceptual framework and trace similar variables (see Chapter 6). The effectiveness of future capacity-building programs could be enhanced by structuring approaches to environmental stewardship based on a careful analysis of traditions of governance, and societal and environmental conditions in a given setting.

Regional programs build on successes achieved on smaller spatial scales. Large marine ecosystems adjoin coastal zones; both are influenced by the rivers, wetlands, and estuaries that transmit the effects of land-based activities to the sea. Hence, goals for the stewardship of marine areas will require efforts beyond traditional sector-by-sector planning and decision-making or experience with ocean and coastal protected areas. Each sector of governance, scaling from local community-based management to national ocean policies and ranging from inland to offshore areas, will be required to establish coordinated and efficient governance.

Convene a Summit on Capacity-Building for Stewardship of Oceans and Coasts

The committee recommends organizing a high-level summit on increasing capacity for stewardship of oceans and coasts to demonstrate political will, to commit to ending fragmentation, to build an agenda for capacity-building that cuts across other programs that address ocean and coastal stewardship issues, and to establish principles and standards for assessing and evaluating program procedures and outcomes. High-level meetings that involve political leaders establish precedents that have the potential to influence policies far into the future.

To date, there has not been the kind of large-scale, high-level summit on capacity-building that will be required to strengthen and coalesce political will among institutional leaders in government, NGOs, and private industry. Capacity-building is usually treated as an ingredient of programmatic efforts on specific topics rather than as a high-priority issue in its own right. For example, capacity-building appears as an element of the Code of Conduct for Responsible Fisheries and of plans for the Global Ocean Observing System. It is addressed in Agenda 21 of the United Nations Conference on Environment and Development (1992), the Millennium Ecosystem Assessments (2005a, b), the United Nations Environment Programme (2006), and so on. In each case, the identification of capacity-building as a critical ingredient is valid, but scattered calls for increased capacity in specific areas make fragmentation of capacity-building efforts all the more likely. Program leaders are likely to place a higher priority on other components of their programs when capacity-building is but one ingredient. Establishment of programs that focus specifically on capacity and solve the problem of fragmentation will require the type of political will that could be generated by a high-level summit. Various types of meetings could serve the purpose, such as a follow-up to the World Summit on Sustainable Development, or the meeting could build, for example, on the United Nations Open-Ended Informal Consultative Process on Oceans and the Law of the Sea.

A high-level summit on capacity-building should involve key leaders who have a stake in stewardship of oceans and coasts, governments, NGOs and intergovernmental organizations, academia, and the private sector. The committee recognizes that meetings of leaders in the field of capacity-building already occur. However, such meetings may not be held at a high enough level to demonstrate political will or end fragmentation and thus might fall short of what will be required to establish an agenda specifically for capacity-building. The committee envisions a high-level summit dedicated to capacity-building that will benefit the broad spectrum of programs working to foster environmental stewardship.

REFERENCES

Agardy, T. 2005. Global marine conservation policy versus site-level implementation: The mismatch of scales and its implications. *Marine Ecology Progress Series* 300:243-248.

Alcala, A.C. 2001. *Marine Reserves in the Philippines: Historical Development, Effects, and Influence on Marine Policy.* Bookmark Publishers, Makati City, Philippines.

Aldo Leopold Leadership Program. 2006. *Aldo Leopold Leadership Program.* [Online]. Available: http://www.leopoldleadership.org [2007, June 12].

Antares Network. 2007. *Antares.* [Online]. Available: http://home.antares.ws/ [2007, September 28].

Brown, L. 2001. *Eco-Economy: Building an Economy for the Earth.* W.W. Norton and Company, New York, New York.

Caldeira, K. and M.E. Wickett. 2003. Anthropogenic carbon and ocean pH. *Nature* 425:365.

Cap-Net. 2004. *Cap-Net: Capacity Building for Integrated Water Resources Management.* [Online]. Available: http://www.cap-net.org/ [2007, June 14].

Carpenter, S.R., R. DeFries, T. Dietz, H.A. Mooney, S. Polasky, W.V. Reid, and R.J. Scholes. 2006. Millennium Ecosystem Assessment: Research needs. *Science* 314:257-258.

Charles, R. and J.C. Vermeiren. 2002. *Towards a Solution for Coastal Disasters in the Caribbean.* Prepared for presentation at the Solutions to Coastal Disasters 2002 conference, organized by the American Society of Civil Engineers, San Diego, California, 24-27 February 2002.

ChloroGIN. 2007. *ChloroGIN Africa.* [Online]. Available: http://chlorogin.org/ [2007, October 22].

Chua, T.E. 2007. *The Dynamics of Integrated Coastal Management: Practical Application in the Sustainable Coastal Developments in East Asia.* Partnerships in Environmental Management for the Seas of East Asia, Quezon City, Philippines.

Chua, T.E. and L.F. Scura (eds.). 1992. *Integrative Framework and Methods for Coastal Area Management.* International Center for Living Aquatic Resources Management Conference Proceedings 37, Manila, Philippines.

Cicin-Sain, B. and R.W. Knecht. 1998. *Integrated Coastal and Ocean Management Concepts and Practices.* Island Press, Washington, DC.

Cities Alliance. 2007. *Cities Alliance: Cities without Slums.* [Online]. Available: http://www.citiesalliance.org/index.html [2007, October 2].

Coast Conservation Department. 1990. *Coastal Zone Management Plan.* Colombo, Sri Lanka.

Coast Conservation Department. 1997. *Revised Coastal Zone Management Plan, Sri Lanka.* Colombo, Sri Lanka.

Coleman, J. 1988. Social capital in the creation of human capital. *American Journal of Sociology* 94 Supplement:S95-S190.

113

Commission on Marine Science, Engineering, and Resources. 1969. *Our Nation and the Sea: A Plan for National Action.* United States Government Printing Office, Washington, DC.

Comunidad y Biodiversidad. 2006. *Marine Reserves Pilot Project at Isla Natividad.* [Online]. Available: http://www.cobi.org.mx/publicaciones/eng_fs_natividad_0630.pdf [2007, June 12].

Crawford, B.R., J.S. Cobb, and C.L. Ming (eds.). 1993. *Educating Coastal Managers: Proceedings of the Rhode Island Workshop.* Coastal Resources Management Project, Coastal Resources Center, University of Rhode Island, Narragansett.

de Fontaubert, A.C. 2001. Legal and political considerations. In *The Status of Natural Resources on the High-Seas,* World Wildlife Fund/The World Conservation Union (IUCN) (eds.). World Wildlife Fund/The World Conservation Union (IUCN), Gland, Switzerland.

Defeo, O. and J.C. Castilla. 2005. More than one bag for the world fishery crisis and keys for co-management successes in selected artisanal Latin American shellfisheries. *Reviews in Fish Biology and Fisheries* 15(3):265-283.

Duda, A.M. 2002. *Monitoring and Evaluation Indicators for GEF International Waters Projects.* Monitoring and Evaluation Working Paper 10, Global Environment Facility, Washington, DC.

Edwards, S.F. 2005. Ownership of multi-attribute fishery resources in large marine ecosystems. In *Sustaining Large Marine Ecosystems: The Human Dimension,* Hennessey T. and J.G. Sutinen (eds.). Elsevier B.V., Amsterdam, Netherlands.

Emerton, L., J. Bishop, and L. Thomas. 2006. *Sustainable Financing of Protected Areas: A Global Review of Challenges and Options.* The World Conservation Union, Gland, Switzerland and Cambridge, UK.

Food and Agriculture Organization of the United Nations. 2007a. *The State of World Fisheries and Aquaculture 2006.* Food and Agriculture Organization, Fisheries and Aquaculture Department, Rome, Italy.

Food and Agriculture Organization of the United Nations. 2007b. *Vessel Monitoring Systems.* [Online]. Available: http://www.fao.org/fi/website/FIRetrieveAction.do?dom=topic&fid=13691 [2007, June 14].

Food and Agriculture Organization of the United Nations. 2007c. *Implementation of the 1995 FAO Code of Conduct for Responsible Fisheries.* [Online]. Available: http://www.fao.org/fi/website/FIRetrieveAction.do?dom=org&xml=CCRF_prog.xml [2007, June 13].

Glasbergen, P. (ed.). 1998. *Co-operative Environmental Governance: Public-Private Agreements as a Policy Strategy.* Kluwer Academic Publishers, Dordrecht, Netherlands.

Hale, L.Z. and E. Kumin. 1992. *Implementing a Coastal Resources Management Policy: The Case of Prohibiting Coral Mining in Sri Lanka.* Coastal Resources Center, University of Rhode Island, Narragansett.

Hanna, S. 1998. Institutions for marine ecosystems: Economic incentives and fishery management. *Ecological Applications* 8(suppl):S170-S174.

Harris, J. 2006. *The Role of Local Governments in Protecting Oceans and Coasts Local Government Capacity Building.* Presentation at National Academies Workshop "Meeting the Challenges of Capacity Building for Managing Oceans and Coasts," Panamá City, Panamá, November 14, 2006.

Hennessey, T. 1994. Governance and adaptive management for estuarine ecosystems: The case of Chesapeake Bay. *Coastal Management* 22:119-145.

Holling, C.S. (ed.). 1978. *Adaptive Environmental Assessment and Management.* John Wiley and Sons, New York, New York.

Hudson Institute. 2006. *The Index of Global Philanthropy.* Hudson Institute, Washington, DC.

Imperial, M. and T. Hennessey. 1993. The evolution of adaptive management for estuarine ecosystems: The National Estuary Program and its precursors. *Ocean and Coastal Management* 20:147-180.

Inter-American Development Bank. 1998. *Sustainable Development Program for Darién (PN-0116: Executive Summary.* [Online]. Available: http://idbdocs.iadb.org/wsdocs/getdocument.aspx?docnum=460850 [2007, October 18].

Inter-American Development Bank. 2004. *Haiti: Bank's Transition Strategy 2005-2006.* [Online]. Available: http://idbdocs.iadb.org/wsdocs/getdocument.aspx?docnum=561125 [2007, September 18].

Intergovernmental Oceanographic Commission. 2002. *Annual Report 2001.* Intergovernmental Oceanographic Commission Annual Reports Series No. 8, United Nations Educational, Scientific and Cultural Organization.

Intergovernmental Panel on Climate Change. 2007. *Climate Change 2007: Climate Change Impacts, Adaptation and Vulnerability—Summary for Policymakers* [Online]. Available: http://www.ipcc.ch/SPM6avr07.pdf [2007, September 20].

International Ocean Institute. 2005. *International Ocean Institute.* [Online]. Available: http://www.ioinst.org/ [2007, June 13].

Joint Group of Experts on Scientific Aspects of Marine Environmental Protection (GESAMP). 1996. The contributions of science to integrated coastal management. *GESAMP Reports and Studies No. 61.* Food and Agriculture Organization of the United Nations, Rome, Italy.

Juda, L. 1999. Considerations in developing a functional approach to the governance of large marine ecosystems. *Ocean Development and International Law* 30(2):89-125.

Juda, L. and T. Hennessey. 2001. Governance profiles and the management and use of large marine ecosystems. *Ocean Development and International Law* 32(1):43-69.

Kates, R.W., W.C. Clark, R. Corell, J.M. Hall, C.C. Jaeger, I. Lowe, J.J. McCarthy, H. Joachim Schellnhuber, B. Bolin, N.M. Dickson, S. Faucheux, G.C. Gallopin, A.Grübler, B.Huntley, J.Jäger, N.S. Jodha, R.E. Kasperson, A. Mabogunje, P. Matson, H. Mooney, B. Moore III, T. O Riordan, and U. Svedin. 2001. Sustainability science. *Science* 292(5517):641-642.

Kimball, L.A. 2001. *International Ocean Governance: Using International Law and Organizations to Manage Marine Resources Sustainability.* The World Conservation Union, Gland, Switzerland and Cambridge, UK.

Lee, K.N. 1993. *Compass and Gyroscope: Integrating Science and Politics for the Environment.* Island Press, Washington, DC.

Leslie, H.M. and A.P. Kinzig. In review. Ecosystem based management: The need for a coupled-social-ecological perspective. *BioScience.*

Levin, S.A., and J. Lubchenco. Submitted. Resilience, robustness, and marine ecosystem-based management. *BioScience.*

Liu, J., T. Dietz, S.R. Carpenter, M. Alberti, C. Folke, E. Moran, A.N. Pell, P. Deadman, T. Kratz, J. Lubchenco, E. Ostrom, Z. Ouyang, W. Provencher, C.L. Redman, S.H. Schneider, and W.W. Taylor. 2007. Complexity of coupled human and natural systems. *Science* 317:1513-1516.

Liu, J., T. Dietz, S.R. Carpenter, C. Folke, M. Alberti, C.L. Redman, S.H. Schneider, E. Ostrom, A.N. Pell, J. Lubchenco, W.W. Taylor, Z. Ouyan, P, Deadman, T. Kratz, and W. Provencher. In press. Coupled human and natural systems. *Ambio.*

Lotze, H.K., H.S. Lenihan, B.J. Bourque, R.H. Bradbury, R.G. Cooke, M.C. Kay, S.M. Kidwell, M.X. Kirby, C.H. Peterson, and J.B.C. Jackson. 2006. Depletion, degradation, and recovery potential of estuaries and coastal seas. *Science* 312:1806-1809.

Lowell, B.L., A. Findlay, and E. Stewart. 2004. Brain strain: Optimising highly skilled migration from developing countries. *Institute for Public Policy Research, Asylum and Migration Working Paper 3.* London, United Kingdom.

Lowry, K. and H.J.M. Wickremeratne. 1989. Coastal area management in Sri Lanka. *Ocean Yearbook* 7:263-293.

Lubchenco, J. 1998. Entering the Century of the Environment: A New Social Contract for Science. *Science* 279:491-497.

Marine Stewardship Council. 2002. *The Marine Stewardship Council.* [Online]. Available: http://www.msc.org [2007, June 13].

Millennium Ecosystem Assessment. 2005a. *Ecosystems and Human Well-Being: Current State and Trends, Volume 1.* World Resources Institute, Washington, DC.

Millennium Ecosystem Assessment. 2005b. *Ecosystem Services and Human Well-Being: Synthesis.* World Resources Institute, Washington, DC.

Millennium Ecosystem Assessment. 2005c. *Ecosystems and Human Well-Being: Scenarios, Volume 2.* World Resources Institute, Washington, DC.

Millennium Ecosystem Assessment. 2005d. *Ecosystems and Human Well-Being: Policy Responses, Volume 3.* World Resources Institute, Washington, DC.

Mizrahi, Y. 2004. *Capacity Enhancement Indicators: Review of the Literature.* WBI Working Papers, World Bank Institute, The World Bank, Washington, DC.

Moberg F. and P. Rönnbäck. 2003. Ecosystem services in the tropical seascape: Ecosystem interactions, substituting technologies, and ecosystem restoration. *Ocean and Coastal Management* 46:27-46.

National Academy of Public Administration. 2006. *Why Foreign Aid to Haiti Failed.* Washington, DC.

National Oceanic and Atmospheric Administration. 2007. *Banco Chinchorro Biosphere Reserve (RBBCH), México.* [Online]. Available: http://effectivempa.noaa.gov/sites/chinchorro.html#overview [2007, June 13].

National Research Council. 1995. *Colleges of Agriculture at the Land Grant Universities: A Profile.* National Academy Press, Washington, DC.

National Research Council. 1999. *Sustaining Marine Fisheries.* National Academy Press, Washington, DC.

National Research Council. 2001. *Marine Protected Areas: Tools for Sustaining Ocean Ecosystems.* National Academy Press, Washington, DC.

National Research Council. 2002. *The Drama of the Commons.* National Academy Press, Washington, DC.

National Research Council. 2004. *Improving the Use of the "Best Scientific Information Available" Standard in Fisheries Management.* The National Academies Press, Washington, DC.

National Research Council. 2005. *Policy Implications of International Graduate Students and Postdoctoral Scholars in the United States.* The National Academies Press, Washington, DC.

National Round Table on the Environment and the Economy. 1998. *Sustainable Strategies for Oceans: A Co-Management Guide.* National Round Table on the Environment and the Economy, Ottawa, Canada.

The Nature Conservancy. 2000. The Five-S Framework for Site Conservation: A Practitioner's Handbook for Site Conservation Planning and Measuring Conservation Success. The Nature Conservancy, Arlington, Virginia.

Network of Aquaculture Centres in Asia-Pacific. 2007. *Network of Aquaculture Centres in Asia-Pacific: About Us.* [Online]. Available: http://www.enaca.org/modules/tinyd1/ [2007, April 27].

Noakes, D.J., L. Fang, K.W. Hipel, and D.M. Kilgour. 2003. An examination of the salmon aquaculture conflict in British Columbia using the graph model for conflict resolution. *Fisheries Management and Ecology* 10(3):123-137.

Office of the Auditor General of Canada. 2000. *Fisheries and Oceans—The Effects of Salmon Farming in British Columbia on the Management of Wild Salmon Stocks.* [Online]. Available: http://www.oag-bvg.gc.ca/domino/reports.nsf/html/0030ce.html [2007, September 18].

Olsen, S.B. 2000. Educating for the governance of coastal ecosystems: The dimensions of the challenge. *Ocean and Coastal Management* 43:331-341.

Olsen, S.B. 2003. Frameworks and indicators for assessing progress in integrated coastal management initiatives. *Ocean and Coastal Management* 46:347-361.

Olsen, S.B., J. Tobey, and L. Hale. 1998. A learning-based approach to coastal management. *Ambio* 27:611-619.

Olsen, S.B., K. Lowry, and J. Tobey. 1999. *A Manual for Assessing Progress in Coastal Management.* Coastal Resources Center, University of Rhode Island, Narragansett.

Olsen, S.B. and P. Christie. 2000. What are we learning from tropical coastal management experiences? *Coastal Zone Management Journal* 28:5-18.

Olsen, S.B. and D. Nickerson. 2003. *The Governance of Coastal Ecosystems at the Regional Scale: An Analysis of the Strategies and Outcomes of Long-Term Programs.* Coastal Resources Center, University of Rhode Island, Narragansett.

Olsen, S.B., J.G. Sutinen, L. Juda, T.M. Hennessey, and T.A. Grigalunas. 2006a. *A Handbook on Governance and Socioeconomics of Large Marine Ecosystems.* Coastal Resources Center, University of Rhode Island, Narragansett.

Olsen, S.B., P.V. Padma, and B. Richter. 2006b. *A Guide to Managing Freshwater Inflows to Estuaries.* The Nature Conservancy, Washington, DC; and Coastal Resources Center, University of Rhode Island, Narragansett.

One Laptop per Child Foundation. 2007. *One Laptop per Child Foundation.* [Online]. Available: http://laptopfoundation.org/en/index.shtml [2007, June 14].

Open Source Initiative. 2007. *Open Source.* [Online]. Available: http://www.opensource.org/ [2007, June 14].

Partnership for Observation of the Global Oceans. 2001. *The São Paulo Declaration.* [Online]. Available: http://www.ocean-partners.org/documents/docSPD.pdf [2007, May 30].

Partnership for Observation of the Global Oceans. 2007. *Training/Education.* [Online]. Available: http://www.ocean-partners.org/Training_Education.htm [2007, October 22].

Partnerships in Environmental Management for the Seas of East Asia. 2006. *Performance Evaluation: Building Partnerships in Environmental Management for the Seas of East Asia (PEMSEA)—Terminal Evaluation Report.* PEMSEA Information Series. Global Environment Facility/United Nations Development Programme/International Maritime Organization Regional Programme on Building Partnerships in Environmental Management for the Seas of East Asia (PEMSEA), Quezon City, Philippines.

Payoyo, P.B. (ed.). 1994. *Ocean Governance: Sustainable Development of the Seas.* United Nations University Press, Tokyo, Japan.

Pew Oceans Commission. 2003. *America's Living Oceans: Charting a Course for Sea Change.* Pew Oceans Commission, Arlington, Virginia.

Pomeroy, R.S., J.E. Parks, and L.M. Watson. 2004. *How is Your MPA Doing? A Guidebook of Natural and Social Indicators for Evaluating Marine Protected Area Management Effectiveness.* The World Conservation Union, Gland, Switzerland and Cambridge, UK.

Pretty, J. 2003. Social capital and the collective management of resources. *Science* 302:1912-1914.

Putman, R.D. 1993. *Making Democracy Work: Civic Traditions in Modern Italy.* Princeton University Press, Princeton, New Jersey.

Putnam, R.D. 2000. *Bowling Alone: The Collapse and Revival of American Communities.* Simon & Schuster, New York, NY.

Raymundo, L.J. 2002. Community-based coastal resources management of Apo Island, Negros Oriental, Philippines: History and lessons learned. *Report from the First ICRAN Regional Workshop on Experience Sharing Between Demonstration and Target Sites in the East Asian Seas (26-27 August 2002).* United Nations Environment Programme, Bangkok, Thailand.

Reece, M.E. 2000. *Global Positioning System: History of Navigation.* [Online]. Available: http://infohost.nmt.edu/~mreece/gps/history.html [2007, June 14].

The Royal Society. 2005. *Ocean Acidification Due to Increasing Atmospheric Carbon Dioxide.* The Clyvedon Press Ltd., Cardiff, UK.

Scientific Committee on Oceanic Research. 2007. *Capacity-Building Activities.* [Online]. Available: http://www.scor-int.org/capacity.htm [2007, September 26].

Sissenwine, M.P. 2007. Environmental science, environmentalism, and governance. *Environmental Conservation* 34(2):21-22.

Smart Growth Network. 2007. *Smart Growth Online.* [Online]. Available: http://www.smartgrowth.org/default.asp [2007, October 2].

Smith, T., B. Fulton, A. Hobday, D. Smith, and P. Shoulder. 2006. *Scientific Tools to Support Practical Implementation of EBFM.* [Online]. Available: http://www.ices06sfms.com/documents/Session%20No%201/1%20Smith.ppt [2007, October 22].

Spergel, B. and M. Moye. 2004. *Financing Marine Conservation: A Menu of Options.* World Wildlife Fund, Center for Conservation Finance, Washington, DC.

Suman, D. 2005. Globalization and development: Using the coastal area of the Darién region of Panamá as a case study. *Occasional Papers on Globalization* 2(5):1-20.

Suman, D. 2007. Development of an integrated coastal management plan for the Gulf of San Miguel and Darién Province, Panamá: Lessons from the experience. *Ocean and Coastal Management* 50(8):634-660.

Sutinen, J. (ed.). 2000. *A Framework for Monitoring and Assessing Socio-Economic and Governance of Large Marine Ecosystems.* National Oceanic and Atmospheric Administration Technical Memorandum NMFS-NE-158.

Swedish Agency for International Development Cooperation. 2007. *Marine and Coastal Research.* [Online]. Available: http://www.sida.se/sida/jsp/sida.jsp?d=673&a=4478 [2007, September 26].

Third World Academy of Sciences. 2004. *Building Scientific Capacity: A TWAS Perspective.* Report of the Third World Academy of Sciences, Trieste, Italy.

U.S. Agency for International Development. 2006. Effective anticorruption approaches: U.S. Agency for International Development (USAID), Office of Democracy and Governance. *Issues of Democracy* 11(12):14-15.

U.S. Commission on Ocean Policy. 2004. *An Ocean Blueprint for the 21st Century.* U.S. Commission on Ocean Policy, Washington, DC.

U.S. Department of State. 2004. *Capacity Building for the Protection and Sustainable Use of Oceans and Coasts.* Proceedings of a Symposium held November 8-9, 2004, in cooperation with the Ocean Studies Board of the National Academies, Washington, DC.

U.S. Geological Survey. 2007. *Geographic Information Systems Poster.* [Online]. Available: http://erg.usgs.gov/isb/pubs/gis_poster/#what [2007, June 13].

United Nations. 2002. *Johannesburg Plan of Implementation of the World Summit on Sustainable Development.* United Nations, New York, New York.

United Nations. 2005. *The Millennium Development Goals Report.* United Nations, New York, New York.

United Nations. 2006. *Train-Sea-Coast Programme: A Cooperative Training Programme in the Field of Coastal and Ocean Management.* [Online]. Available: http://www.un.org/Depts/los/tsc_new/TSC_about.pdf [2007, September 26].

United Nations Conference on Environment and Development. 1992. *Agenda 21: A Programme for Action for Sustainable Development.* United Nations, New York, New York.

United Nations Development Programme. 2005. *Survivors of the Tsunami: One Year Later.* United Nations Development Programme, New York, New York.

United Nations Division for Sustainable Development. 2005. *Sustainable Development Topics.* [Online]. Available: http://www.un.org/esa/sustdev/sdissues/capacity_building/capacity.htm [2007, June 14].

United Nations Environment Programme. 1995. Guidelines for Integrated Management of Coastal and Marine Areas with Special Reference to the Mediterranean Basin. UNEP Regional Seas Reports and Studies No. 161. United Nations Environment Programme, Nairobi, Kenya.

United Nations Environment Programme. 2004. *Inventory of UNEP Capacity-Building and Technology Support Activities.* High-Level Open-Ended Intergovernmental Working Group on an Intergovernmental Strategic Plan for Technology Support and Capacity-Building. Second session, Nairobi, September 2-4, 2004.

United Nations Environment Programme. 2005. *Regional Seas Programme.* [Online]. Available: http://www.unep.org/regionalseas/ [2007, June 14].

United Nations Environment Programme. 2006. *Marine and Coastal Ecosystems and Human Well-Being: A Synthesis Report Based on the Findings of the Millennium Ecosystem Assessment.* United Nations Environment Programme, Nairobi, Kenya.

United Nations Environment Programme/Global Programme of Action for the Protection of the Marine Environment from Land-Based Activities. 2006. *Ecosystem-Based Management: Markers for Assessing Progress.* The Hague, Netherlands.

United Nations Human Settlement Programme. 2007. *Sustainable Cities Programme.* [Online]. Available: http://www.unhabitat.org/content.asp?typeid=19&catid=540&cid=5025 [2007, October 2].

Valiela, I., J.L. Bowen, and J.K. York. 2001. Mangrove forests: One of the world's threatened major tropical environments. *BioScience* 51(10):807-815.

Villa, F., L. Tunesi, and T. Agardy. 2002. Zoning marine protected areas through spatial multiple-criteria analysis: The case of the Asinara Island National Marine Reserve of Italy. *Conservation Biology* 16(2):515-526.

Vitousek, P.M., H.A. Mooney, J. Lubchenco, and J.M. Melillo. 1997. Human domination of Earth's ecosystem. *Science* 277:494-499.

Walker, B. and D. Salt. 2006. *Resilience Thinking.* Island Press, Washington, DC.

Wallis, P. and O. Flaaten. 2000. *Fisheries Management Costs: Concepts and Studies.* Organisation for Economic Co-operation and Development, Paris, France.

Walters, C. 1986. *Adaptive Management and Renewable Resources.* Macmillan, New York, New York.

Ward, T., E. Hegerl, D. Tarte, and K. Short. 2002. *Ecosystem-Based Management of Marine Fisheries: Policy Proposals and Operational Guidance for Ecosystem-Based Management of Marine Capture Fisheries.* World Wildlife Fund, Sydney, Australia.

Wilburn, S.M., J. Tobey, J. Hepp, S. Olsen, and B. Costa-Pierce. 2007. *Sea Grant in Latin America: Adapting the U.S. Sea Grant model of linked applied research, extension, and education to a Latin American context—Is there a fit? Marine Policy* 31(3):229-238.

The World Bank. 2005. *Capacity Building in Africa: An OED Evaluation of World Bank Support.* World Bank Operations Evaluation Department, Washington, DC.

The World Bank. 2006. *Scaling Up Marine Management: The Role of Marine Protected Areas.* [Online]. Available: http://siteresources.worldbank.org/INTCMM/Publications/21108865/Final_Printed_Version_MPA_ESW.pdf. [2007, June 13].

The World Conservation Union. 2000. *Financing Protected Areas: Guidelines for Protected Area Managers—Financing Protected Areas Task Force of the World Commission on Protected Areas (WCPA) of IUCN, in collaboration with the Economics Unit of IUCN,* Phillips, A. (ed.). Best Practice Protected Area Guidelines Series No. 5, The World Conservation Union, Gland, Switzerland.

World Ocean Observatory. 2007. *World Ocean Observatory: A Forum for Ocean Affairs.* [Online]. Available: http://www.thew2o.net/ [2007, June 14].

World Wildlife Fund. 2004. *Are Protected Areas Working?: An Analysis of Forest Protected Areas by WWF.* World Wildlife Fund International, Gland, Switzerland.

Young, O.R.. 1989. *International Cooperation: Building Regimes for Natural Resources and the Environment.* Cornell University Press, Ithaca, New York.

APPENDIXES

A
COMMITTEE AND STAFF BIOGRAPHIES

COMMITTEE

Mary (Missy) H. Feeley (*Cochair*) is chief geoscientist in the Technical Department with the ExxonMobil Exploration Company. She earned a PhD in oceanography from Texas A&M University in 1984. At ExxonMobil, she has been involved in many capacity-building activities in Africa, Asia, and Europe. Her responsibilities include advising senior ExxonMobil Upstream management on strategic geoscience matters and identifying global geoscience opportunities for ExxonMobil. She is also responsible for linking business-unit needs to the geophysical research program in ExxonMobil's Upstream Research Company. Dr. Feeley is a member of the American Association of Petroleum Geologists, the Society of Exploration Geophysicists, and the American Geophysical Union. She is also a member of the Ocean Studies Board.

Silvio Pantoja (*Cochair*) is a professor of oceanography at the University of Concepción in Chile, where he is responsible for a funded university program that aims to build capacity in the ocean sciences. He earned a PhD in oceanography from the State University of New York at Stony Brook. He was a postdoctoral scholar in the Department of Marine Chemistry and Geochemistry at the Woods Hole Oceanographic Institution and a Fulbright Visiting Scientist at the Scripps Institution of Oceanography. His research interests include biogeochemical cycling of organic matter in the marine environment, pathways of degradation and preservation of organic matter, organic biomarkers and stable isotopes as tracers of past and present oceanographic processes, and use of stable isotopes as tracers of sources and diagenetic alteration of organic molecules. He is the chair in oceanography at the Intergovernmental Oceanographic Commission of the United Nations Educational, Scientific, and Cultural Organization.

Tundi Agardy is the founder and executive director of Sound Seas, which works to promote effective marine conservation by using both science and sociology and works as the interface between public policy and community-based conservation. She earned a PhD in biological sciences from the University of Rhode Island in 1987. She has worked globally, specializing in marine protected areas and coastal management, and was formerly employed as a senior scientist at the World Wildlife Fund and senior director of the Global Marine Program for Conservation International.

Juan Carlos Castilla is a professor at the Center for Advanced Studies in Ecology and Biodiversity at the Pontificia Universidad Católica de Chile. He earned a PhD in marine biology from the University College of North Wales in 1970. His research interests include rocky intertidal community structure and dynamics, the ecological role of keystone species, coastal conservation, and comanagement of small-scale benthic resources in Chile and Latin America. He contributed to the establishment and development of the Marine Coastal Station in Las Cruces, Chile, and of marine protected areas along the coast of Chile. Dr. Castilla is a member of the Third World Academy of Sciences, the Chilean Academy of Sciences, and the U.S. National Academy of Sciences.

Stephen Farber is a professor of economics in the Graduate School of Public and International Affairs at the University of Pittsburgh and director of the Environmental Decision Support Program and the Public and Urban Affairs Program. Dr. Farber earned a PhD in economics from Vanderbilt University in 1973. His research focuses on the economics of ecosystems, particularly the economic valuation of ecosystems and their services. He has served as a consultant and advisory board member for coastal management, watershed management, regional sewer and waste treatment, and urban sustainability programs.

Indumathie Hewawasam is a senior environmental specialist at the World Bank. She earned a law degree in Sri Lanka and practiced as an attorney at the official bar in Sri Lanka in the late 1970s and early 1980s. She earned an MA in international development and environmental management from American University, Washington, DC, and a PhD in marine policy from the University of Delaware. Her expertise is in coastal and marine resource management, community-based natural resources management, and poverty and sustainable development in Africa. Among her contributions to marine policy at the World

Bank are the publications *Managing the Marine and Coastal Environment of Sub-Saharan Africa: Strategic Directions* and *Blueprint 2050: Sustaining the Marine Environment in Mainland Tanzania and Zanzibar.* The last decade of her work at the World Bank has focused on assisting developing nations with policy and institutional development toward sound management of their exclusive economic zones, territorial waters, coastal zones, sustainable livelihoods, and poverty reduction through sound natural resources management. She also advises on the World Bank's Tsunami Rehabilitation Program in South India. Outside the World Bank, Dr. Hewawasam serves on the Steering Committee of the Global Forum on Oceans, Coasts, and Islands and on the Advisory Committee of the Pew Fellows in Marine Conservation.

Joanna Ibrahim is a lecturer in coastal engineering and coastal zone management at the University of the West Indies, where she is establishing an active coastal research unit in the Civil and Environmental Engineering Department. She earned a PhD from the University of Plymouth (UK) in 1998. Her research interests include coastal zone management and process issues, such as sediment transport, beach morphodynamics, and natural hazards vulnerability.

Jane Lubchenco is a Distinguished Professor of Zoology and Wayne and Gladys Valley Professor of Marine Biology at Oregon State University. She earned a PhD in ecology from Harvard University in 1975. Her research interests include public understanding of science, marine conservation, ecosystem services, ecological consequences of global change, biodiversity, and sustainable ecological systems. Dr. Lubchenco is a member of the National Academy of Sciences.

Bonnie McCay is a Board of Governors Distinguished Service Professor in the Department of Human Ecology at Cook College of Rutgers University. She earned a PhD in anthropology from Columbia University in 1976. Her research interests include the socioeconomic, cultural, and political dimensions of marine fisheries and fisheries management and community responses to industrialization related to offshore wind energy. She is vice chair of the Federal Advisory Committee on Marine Protected Areas.

Nyawira Muthiga is a conservation scientist at the Wildlife Conservation Society and coordinates its marine programs in the Western Indian Ocean, including projects in Kenya, Tanzania, Mauritius, the Seychelles, and Madagascar. She earned a PhD in zoology from the University of Nairobi in 1996. Dr. Muthiga's research focuses on monitoring coral reefs in protected and unprotected areas along the Kenyan coast, socioeconomic perspectives of coastal resource use, and the process of integrated coastal management. Her work involves direction of management planning, research and monitoring, and training programs in the management and use of coastal and marine wetlands, awareness of the value of wetlands, and conservation geared to the sustainable management of these ecosystems. She received the National Geographic/Buffett Award for achievements in conservation and the Kenyan resident's award, the Order of the Grand Warrior. Dr. Muthiga is also president of the Western Indian Ocean Marine Science Association.

Stephen Olsen is director of the Coastal Resources Center at the University of Rhode Island. He earned an MS in biological oceanography from the University of Rhode Island in 1970. His research focuses on capacity-building to address anthropogenic changes in coastal ecosystems, governance of coastal ecosystems, and formulation, application, and refinement of coastal management initiatives in developed and developing nations. He has worked on developing a learning-based, issue-driven approach to the management of coastal ecosystems through long-term programs in Latin America, East Africa, and Southeast Asia. Mr. Olsen has worked with the World Bank, the Inter-American Development Bank, the United Nations Development Programme, and the U.S. Agency for International Development on their coastal management initiatives in a growing number of countries. He has also been working to formulate a common method for learning from coastal management experience.

Shubha Sathyendranath is the executive director of the Partnership for Observation of the Global Oceans (POGO) at the Bedford Institute of Oceanography in Dartmouth, Nova Scotia, Canada. Her work at POGO focuses on implementation issues, such as technical compatibility among observing networks, shared use of infrastructure, and public outreach and capacity-building; and she has been a participant in international forums on capacity-building. She earned a PhD in optical oceanography from the Université Pierre & Marie Curie in Paris, France, in 1981. Her research interests include marine optics, remote sensing, and international coordination, cooperation, and research.

Michael Sissenwine is an independent marine-science consultant and Visiting Scholar of the Woods Hole Oceanographic Institution. He served as director of scientific programs and chief science adviser for the National Marine Fisheries Service for three years, until June 2005, and previously as director of the Northeast Fisheries Science Center for six years. Dr. Sissenwine has been active in capacity-building projects on several continents; his research interests include ecosystem dynamics, fisheries oceanography, resource assessments, and fishery management theory and case studies. He has served as president of the International Council for the Exploration of the Sea, as U.S. delegate to the Pacific Science Association, as a member of the Commission on Fisheries Resources of the World Humanity Action Trust in the UK, as chair of the Advisory Committee on Fishery Research for the United Nations Food and Agriculture Organization, and as a member of the Ocean Studies Board of the National Research Council.

Daniel Suman is a professor in the Division of Marine Affairs and Policy in the Rosenstiel School of Marine and Atmospheric Science at the University of Miami and is an adjunct professor in the School of Law. He earned a PhD in chemical oceanography from the Scripps Institution of Oceanography and an MEd in international education, an MA in comparative education, and a certificate in Latin American studies from Columbia University. In addition, Dr. Suman has a JD and a certificate in environmental law from the University of California, Berkeley. His main research interests are the adaptability of the fishing sectors in Chile, Perú, and Ecuador to El Niño-Southern Oscillation climate variability, including the response of industrial fishing companies and labor unions, artisanal fishing unions, and government regulators to environmental uncertainty; mangrove management in Latin American and Caribbean countries; coastal area management and planning; and conflict resolution in marine protected areas. Much of his research focuses on Central and South America, particularly Panamá. He was a member of the Ocean Studies Board and is a member of the World Conservation Union's Commission on Environmental Law, and the Federal Advisory Committee on Marine Protected Areas.

Giselle Tamayo is a professor of natural products and organic synthesis at the University of Costa Rica. She earned a PhD from the Technical University of Berlin in 1989. She also leads a group in genetic and genomic research and is involved in international efforts to develop appropriate ways for developed and developing countries to work together to sample and study genetic resources. She is a scientific adviser to Costa Rica's diplomatic teams that develop access and benefit treaties, such as the United Nations Convention on Biological Diversity.

STAFF

Susan Roberts became the director of the Ocean Studies Board in April 2004. She received her PhD in marine biology from the Scripps Institution of Oceanography. Dr. Roberts's research experience has included fish muscle physiology and biochemistry, marine bacterial symbioses, and developmental cell biology. She has directed a number of studies for the Ocean Studies Board, including those which led to the reports *A Review of the Ocean*

Research Priorities Plan and Implementation Strategy (2007); *Mitigating Shore Erosion Along Sheltered Coasts* (2007); *Nonnative Oysters in the Chesapeake Bay* (2004); *Decline of the Steller Sea Lion in Alaskan Waters: Untangling Food Webs and Fishing Nets* (2003); *Effects of Trawling and Dredging on Seafloor Habitat* (2002); *Marine Protected Areas: Tools for Sustaining Ocean Ecosystems* (2001); *Under the Weather: Climate, Ecosystems, and Infectious Disease* (2001); *Bridging Boundaries Through Regional Marine Research* (2000); and *From Monsoons to Microbes: Understanding the Ocean's Role in Human Health* (1999). Dr. Roberts specializes in the science and management of living marine resources.

Frank R. Hall joined the Ocean Studies Board as a program officer in January 2006 and served as the study director for the committee through July 2007. He received his PhD in oceanography from the University of Rhode Island in 1991. His dissertation research involved quaternary paleoceanographic reconstructions of the high-latitude Atlantic and Arctic Oceans. In 1994, he was awarded a Ford Foundation Postdoctoral Fellowship to study at the Institute for Arctic and Alpine Research. In 1998, Dr. Hall joined the faculty at the University of New Orleans as a geoscience educator, focusing on the preparation of preservice and inservice grades K–12 science teachers. Before joining the Ocean Studies Board, he served as a program officer in the Division of Elementary, Secondary, and Informal Education at the National Science Foundation.

Jodi Bostrom is a research associate with the Ocean Studies Board. She earned an MS in environmental science from American University in 2006 and a BS in zoology from the University of Wisconsin-Madison in 1998. Since starting with the board in May 1999, Ms. Bostrom has worked on several studies pertaining to coastal restoration, fisheries, marine mammals, nutrient overenrichment, ocean exploration, and capacity-building.

B

PANAMÁ CONFERENCE 2006: ARE WE MEETING THE CHALLENGES OF CAPACITY-BUILDING FOR MANAGING OCEANS AND COASTS?

AGENDA

Smithsonian Tropical Research Institute
Balboa, Ancón
Panamá
November 13-14, 2006

Monday, November 13

8:30 AM   Introduction of the Co-Chairs: Missy Feeley and Silvio Pantoja—*Frank Hall*
Welcoming Remarks—*Missy Feeley* and *Silvio Pantoja*
Introduction of Ira Rubinoff, STRI Director—*Missy Feeley* and *Silvio Pantoja*
Welcoming Remarks—*Ira Rubinoff*
Introduction of Gregory Symmes, Division of Earth and Life Studies—*Missy Feeley* and *Silvio Pantoja*

Welcoming Remarks—Gregory Symmes

Introduction of the Committee Members—*Missy Feeley* and *Silvio Pantoja*

SESSION 1	**PANAMÁ: A MODEL OF EXPERIENCE IN BUILDING CAPACITY**—*Daniel Suman* (moderator)
9:15 AM	*Julio Escobar* (National Secretariat for Science, Technology, and Innovation, Panamá)
9:30 AM	*Orlando Osorio* (Program for the Sustainable Development of Darién Province, Panamá)
9:45 AM	*Glenis Binns* (Program for the Sustainable Development of Bocas del Toro Province, Panamá)
10:00 AM	*Líder Sucre* (National Association for the Conservation of Nature, Panamá)
10:15 AM	*Rosa Montañez* (Ramsar Regional Center for Wetland Training and Research in the Western Hemisphere, Panamá)
10:30 AM	Break
11:00 AM	*Charlotte Elton* (Panamanian Center for Research and Social Action)
11:15 AM	*Zuleika Pinzón* (Nature Foundation, Panamá)
11:30 AM	Roundtable Discussion/Question and Answer
1:00 PM	Lunch available in the STRI cafeteria.
SESSION 2	**PERSPECTIVES FROM THOSE WHO SUPPORT AND FUND CAPACITY BUILDING EFFORTS**—*Stephen Farber* and *Silvio Pantoja* (moderators)
2:00 PM	*Henrik Franklin* (Inter-American Development Bank, United States)
2:15 PM	*Takashi Ito* (Nippon Foundation, Japan)
2:30 PM	*Richard Volk* (U.S. Agency for International Development)
2:45 PM	Roundtable Discussion/Question and Answer
3:30 PM	Break
SESSION 3	**PERSPECTIVES FROM THOSE WORKING IN AND WITH THE COMMUNITIES**—*Tundi Agardy* and *Michael Sissenwine* (moderators)
4:00 PM	*Peter Burbridge* (University of Newcastle, England)

4:15 PM	*Stella Maris Vallejo* (United Nations Train-Sea-Coast Program [retired], Portugal)
4:30 PM	Roundtable Discussion/Question and Answer
6:00 PM	Reception

Tuesday, November 14

SESSION 4	**PERSPECTIVES FROM GOVERNMENTAL ENTITIES**—*Juan Carlos Castilla* and *Shubha Sathyendranath* (moderators)
8:30 AM	*Lorna Inniss* (Coastal Zone Management Unit, Barbados)
8:45 AM	*Mirei Endara de Heras* (CALI Foundation, Panamá)
9:00 AM	*Genevieve Brighouse* (American Samoa Coastal Management Program)
9:15 AM	Roundtable Discussion/Question and Answer
10:30 AM	Break
SESSION 5	**PERSPECTIVES FROM THOSE WHO DESIGN AND ADMINISTER PROJECTS AND PROGRAMS**—*Bonnie McCay* and *Frank Hall* (moderators)
11:00 AM	*Jeremy Harris* (former Mayor of Honolulu, Hawai'i, United States)
11:15 AM	*Barry Costa-Pierce* (University of Rhode Island, United States)
11:30 AM	*Loke Ming Chou* (National University of Singapore)
11:45 AM	*Patrick Christie* (University of Washington, United States)
12:00 PM	Roundtable Discussion/Question and Answer
1:00 PM	Lunch available in the STRI cafeteria.
2:00 PM	Charge to the Breakout Groups: "Next Steps"—*Missy Feeley* and *Silvio Pantoja*
2:30 PM	Breakout Groups Group 1: Sustainability Beyond the End of Funding—*Jane Lubchenco* and *Frank Hall* (moderators) Group 2: Erecting and Maintaining Global Networks—*Stephen Olsen* and *Daniel Suman* (moderators)

Group 3: Technology, Infrastructure, and Education: Necessary Components to Build Capacity—*Joanna Ibrahim* and *Missy Feeley* (moderators)

4:00 PM   Break

4:30 PM   Continue Breakout Groups

5:30 PM   Breakout Groups End (Reassemble in Auditorium)

5:45 PM   Results of Breakout Groups
              Group 1: *Jane Lubchenco* and *Frank Hall*
              Group 2: *Stephen Olsen* and *Daniel Suman*
              Group 3: *Joanna Ibrahim* and *Missy Feeley*

6:15 PM   Adjournment—*Missy Feeley* and *Silvio Pantoja*

C
MAJOR CHANGES IN CAPACITY-BUILDING SINCE 1969

The evolution of international capacity-building for ocean and coastal management can be divided into three periods. The first extends over 22 years, from the release of the "Stratton Commission" report (Commission on Marine Science, Engineering, and Resources, 1969) in the United States in 1969 until the United Nations Conference on Environment and Development (UNCED) in 1992 (United Nations Conference on Environment and Development, 1992). The second extends over the decade from UNCED to the World Summit on Sustainable Development (WSSD) in 2002 (United Nations, 2002). The third period begins with the WSSD and continues today.

THE FIRST PERIOD

The first period began in the United States with the release of the seminal report *Our Nation and the Sea* (Commission on Marine Science, Engineering, and Resources, 1969). That report promoted a forward-looking and comprehensive approach to ocean and coastal management that gave rise in the following decade to federal legislation that redefined research, planning, and decision-making for coastal development and restoration; the management of fisheries; and the allocation of ocean space and resources. The Coastal Zone Management Act of 1972, a response to *Our Nation and the Sea*, launched an innovative program featuring incentives for state-federal partnerships designed to address high-priority issues raised by the intensification of human activity along coastlines and the degraded condition of the Great Lakes and many estuaries. In addition, the Fishery

Conservation and Management Act[1] of 1976 initiated a system for science-based management of marine fisheries in an expanded zone of jurisdiction (3–200 nautical miles) through regional councils that involved state, federal, industry, and other stakeholder representation.

During the first period, the Law of the Sea Conferences became a forum for international discussion on how the resources of the oceans should be allocated and their exploitation regulated. The issue of varied claims of territorial waters by different countries (for example, the 3-nautical-mile limit, the 12-nautical-mile territorial limit, and the 200-nautical-mile-limit) was raised in the United Nations, and in 1973 the Third United Nations Conference on the Law of the Sea was convened in New York to write a new treaty covering the oceans. The conference concluded in 1982, and over 160 nations participated. In addition to its provisions defining ocean boundaries, the convention established obligations for safeguarding the marine environment and protecting freedom of scientific research on the high seas and created a legal regime for controlling mineral resources exploration in deep seabed areas beyond national jurisdiction.

Capacity-building in support of the new approaches to ocean and coastal management was expressed by large investments in research, much of it conducted by universities, and by encouragement and development of new curricula on ocean and coastal management topics. A central feature of the capacity-building movement was the U.S. National Sea Grant College Program, which was modeled on the success of the U.S. Land-Grant University System in transforming U.S. agriculture and land-use practices. The National Sea Grant College Program was designed to build capacity in society to manage and responsibly use the nation's ocean and coastal resources through university-based programs that combined education, research, and extension in a pragmatic, issue-driven response to issues identified in close consultation with local stakeholders.

Among the international multilateral programs, the United Nations Development Programme (UNDP) was one of the first to sponsor workshops and training courses designed primarily to raise awareness of the importance of oceans and coasts and the emerging approaches to their management. The Food and Agriculture Organization of the United Nations (FAO) sponsored workshops and training for enhanced management of the newly expanded zones of extended fisheries jurisdiction throughout the world and set up regional bodies set up for high seas and international fisheries.

In the 1970s and 1980s, the United States developed an approach to coastal management that was distinguished by the following characteristics:

---

[1]Fishery Conservation and Management Act 16 USC 1801-1882, April 13, 1976, as amended 1978-1980, 1982-1984, 1986-1990, 1992-1994, 1996, and 2007.

- A governance system that differentiated roles and responsibilities in a tiered system composed of municipal (or county), state, and federal governments.
- An issue-driven approach that required addressing an expanding list of topics that were found to be in the national interest and issues of concern to individual states or municipalities.
- A set of rules (different from legislation or policies) for involving the public and all affected parties in the coastal planning and decision-making process.
- A management system that was based on policies and regulations that defined how specific activities were to be conducted, differentiating where various types of activities were to be permitted through zoning and where development was strictly limited or prohibited through the designation of protected areas.

Beginning in 1984, the U.S. Agency for International Development (USAID) sponsored a bilateral international program that worked to adapt what had been learned from the U.S. experience in coastal zone management (CZM) to similar issues in developing nations. In that and later programs and projects, capacity-building to impart the necessary knowledge and skills was a central feature. The USAID-sponsored programs developed many of the first generation of materials designed for use in international coastal management training events to guide the teams charged with developing the methods and institutional frameworks for the management of ocean and coastal space, resources, and activities.

In the initial phase, the most developed models for coastal management programs were the U.S. state CZM programs. The CZM model divided the evolution of a management program into two distinct phases: an initial planning phase and, if that was successful, federal approval of a state program that met an explicit set of standards designed to certify that the necessary capacity had been assembled to implement a program that addressed important coastal issues. It soon became apparent that capacity-building designed to replicate that model was often inappropriate in settings where governments are weak and regulatory approaches have little impact. However, the separation of activities into an initial phase devoted largely to issue analysis and planning and a follow-up phase involving the implementation of a formally sanctioned program shaped the investments; most of the effort was directed at the initial phase.

THE SECOND PERIOD

The second period of capacity-building for ocean and coastal management began with UNCED, held in Rio de Janeiro in 1992. The so-called Earth Summit put forward integrated coastal management (ICM) as the recommended approach for managing the world's coastal regions. The goal was to have ICM programs in place in every coastal nation by 2000. Chapter 17 of UNCED Agenda 21 (United Nations Conference on Environment and Development, 1992) detailed the ICM approach and its defining features.[2] The text was influenced by the U.S. experience and by emerging coastal management programs in a number of developing nations, including Sri Lanka, Barbados, Ecuador, and the Philippines. However, unlike the CZM program in the United States, there were no standards for marking the transition between planning and implementation, no incentives for countries to engage, and no explicit agenda or strategy for building the necessary capacity.

UNCED triggered major bilateral and multilateral investments in ICM projects. With few exceptions, they were designed as short-term investments to help nations to progress through the initial planning phase of an ICM program under the assumption that individual countries would secure the funds to implement a formally adopted ICM program. Capacity-building was invariably one of the components of each project; however, such capacity-building was designed by international funders to meet the immediate needs of short-term projects with short-term courses, mentoring, and study tours selected to complement the activities undertaken by the project in a given period. Because establishing an ICM program on a national scale in the span of the usual five- to ten-year project was usually judged infeasible, many of the projects opted to focus their efforts on pilot-scale demonstrations. It was expected that the capacity generated through an intensive effort in a constrained area would catalyze other efforts in the same country and that the effects would scale up.

During the second period, capacity-building in developing nations was delivered primarily through short-term training and learning-by-doing. By 1993, UNDP had launched its Train-Sea-Coast Program (United Nations, 2006); three years later, the University of Rhode Island was offering its Summer Institute in Coastal Management. Those and other short courses were designed as sets of modules that introduced participants associated with coastal management projects to the fundamental issues and processes of cross-sectoral management. In addition, the International Ocean Institute, based in Malta, established centers at several universities in different world regions and was offering short courses on

---

[2]Other parts of Chapter 17 concern (1) the protection of marine environmental, (2) the sustainable use and conservation of marine living resources both of the high seas and under national jurisdiction, (3) the strengthening of international cooperation and coordination, (4) the sustainable development of small islands, and (5) critical uncertainties in the management of the marine environment and climate change. Those and other provisions of Agenda 21 were presented as a plan to restore heavily depleted stocks by 2015 through a broad approach that included restrictions on fishing, the use of marine protected areas, and stepped-up enforcement (United Nations Conference on Environment and Development, 1992).

ocean law and ocean management. In the same period, the University of Newcastle in England launched its master's degree program in tropical coastal management.

Most projects elected to invest in short-term training rather than degree programs because the years of absence of a person moving toward a degree incurred substantial costs with little immediate benefit to the project. The absence of one or more of the most gifted members of a project team could be a major detriment to a short-term project. Those who did attend degree programs were self-funded or assisted by their home country governments.

The practice of building capacity through short-term courses and on-the-job training proved to have several weaknesses. For the trainee, the multiple, widely advertised training courses available resulted in ad hoc and fragmented infusions with little coherence and often little direct relevance to the trainee's high-priority needs and interests. At the same time, the focus of many courses was on single aspects of the technical parts of management, such as geographic information systems or impact assessment, or on a single topic, such as public education or coastal erosion. There was little if any sequencing of courses. Concurrently, the personnel who attended training courses were mostly middle-level professionals; those responsible for the overall direction of a program could rarely find the time to attend, and there was little incentive for them to do so, inasmuch as the technical nature of the material presented was often not relevant to their needs and interests.

The central features of ICM, the ability to link across sectors and the negotiation of a program of policies and actions involving several competing agencies of government or sectors, could be addressed only in the longer, two- to four-week courses. The emphasis of training during the second period was on the planning phase of coastal management—selecting issues, setting boundaries, building constituencies for a program, and selecting management instruments. The broader challenges of program implementation and sustained funding were not focused on.[3] From the perspective of an emerging professional, participation in a series of disconnected training courses delivered in a variety of pedagogic styles without an overarching design or set of standards provided an awareness of some—primarily technical—aspects of ocean and coastal management practice but not a coherent foundation of its defining features.

Toward the end of the second period, surveys of program participants and needs assessments revealed the need to strengthen the skills of professionals in such topics as

---

[3]One survey of the "flurry of activity in training and teaching" in coastal and ocean management triggered by UNCED (Cicin-Sain and Knecht, 1998) concludes that "a growing segment of these opportunities are *ad hoc* in nature, not part of a larger, more coherent whole. We suggest that efforts be made to focus training opportunities at regional centers and to create networks among training institutions. Greater use of joint programs should be encouraged."

proposal preparation, making presentations, and conflict management, and those have become the focus of capacity-building efforts.

THE THIRD PERIOD

The third, and current, period began with the 2002 Johannesburg WSSD (United Nations, 2002). It was designed to take stock of the progress that had been made in the decade after UNCED and to define global priorities through Millennium Development Goals (United Nations, 2005) with specific targets and timetables; the United Nations Division for Sustainable Development is scheduled to review progress on the goals by 2014–2015. The modest progress in establishing permanent and effective ICM programs was noted at WSSD; because the condition of ocean and coastal resources continued to decline, the scale of integrating forms of management had to be increased. The ecosystem approach to management and the need to develop links between planning and decision-making in watersheds and planning and decision-making along coastlines and in large marine ecosystems were recognized.

WSSD emphasized capacity-building, particularly in Chapter 37 of Agenda 21 (United Nations Conference on Environment and Development, 1992), which focused on national mechanisms and international cooperation in developing countries as necessary means of attaining desired outcomes. Importance was attached to defining country needs and priorities in sustainable development through a continuing participatory process and thereby strengthening human-resources and institutional capabilities (United Nations Division for Sustainable Development, 2005).

In many newly independent former colonies, efforts were made to support and enhance existing national and regional marine-science institutions and to facilitate marine resources development through a regional approach. That approach was difficult to sustain in many areas. In East Africa, for example, the East African Community breakup in 1977 forced capacity-building to the national level, including the Institute for Marine Sciences in Zanzibar. However, this period saw a decline in marine science and fisheries research, which coincided with evidence of decline in ocean and coastal ecosystems and intensified human activities. The situation eventually led to focused efforts by the Swedish development-assistance agency, the Swedish Agency for International Development Cooperation, in marine-science training, and by USAID through the Coastal Resources Center of the University of Rhode Island to create the capacity and political will for coastal management.

In the current period, an initial threshold of capacity and experience in ocean and coastal management is present in each region. Many professionals have participated in one or more projects, and the Internet has greatly facilitated access to materials and communication between interested parties. Regionally, universities are developing courses and degree programs in integrated forms of management. Programs are now attracting more students from Europe and North America than from developing countries; an example is the master's degree program in tropical coastal management at the University of Newcastle. Furthermore, students who a decade ago would have had to move to Europe or North America to earn an advanced degree in coastal management or marine affairs can now select from a number of programs in their home regions. This emerging capacity to educate professionals for careers in integrated management, combined with the presence of a growing cadre of experienced professionals in each region, offers important opportunities to rethink how the increasingly urgent need to build capacity to manage oceans and coasts can be met in coming decades.

D
ACRONYMS

AHP	analytical hierarchy processes
Cap-Net	Capacity Building for Integrated Water Resources Management
CEASPA	Centro de Estudios y Acción Social Panameño (Center for Panamanian Research and Social Action)
COBI	Comunidad y Biodiversidad (Community and Biodiversity)
CRC	Coastal Resources Center
CZM	coastal zone management
EBM	ecosystem-based management
FAO	Food and Agriculture Organization of the United Nations
GEF	Global Environment Facility
GEOSS	Global Earth Observation System of Systems
GIS	geographic information system
GPS	global positioning system
ICM	integrated coastal management
IOC	Intergovernmental Oceanographic Commission
IOCCG	International Ocean-Colour Coordinating Group
IOI	International Ocean Institute
IUCN	World Conservation Union

JAMSTEC	Japan Agency for Marine-Earth Science and Technology
LME	large marine ecosystem
MACEMP	Marine and Coastal Environmental Management Project
MEABR	management and exploitation area for benthic resources
MPA	marine protected area
MSC	Marine Stewardship Council
NACA	Network of Aquaculture Centres in Asia-Pacific
NGO	nongovernmental organization
OECD	Organisation for Economic Co-operation and Development
PDSD	Program for the Sustainable Development of Darién
PEMSEA	Partnerships in Environmental Management for the Seas of East Asia
PISCO	Partnership for Interdisciplinary Studies of Coastal Oceans
POGO	Partnership for Observation of the Global Oceans
SCOR	Scientific Committee on Oceanic Research
UNCED	United Nations Conference on Environment and Development
UNDP	United Nations Development Programme
UNEP	United Nations Environment Programme
USAID	U.S. Agency for International Development
VMS	vessel monitoring system
WSSD	World Summit on Sustainable Development
WWF	World Wildlife Fund